SPACE ROCKS

WOMEN'S ADVENTURES IN SCIENCE

SPACE ROCKS
the story of planetary geologist
ADRIANA OCAMPO

by Lorraine Jean Hopping

Franklin Watts
A Division of Scholastic Inc.
New York • Toronto • London • Auckland • Sydney
Mexico City • New Delhi • Hong Kong
Danbury, Connecticut

Joseph Henry Press
Washington, D.C.

Author's Acknowledgments

I would like to thank Adriana Ocampo and her family for sharing their inspirational story of hard work, persistence, and always following your dreams. Many of Adriana's friends and colleagues graciously agreed to interviews or made other contributions: Walter Alvarez, Lu Coffing, Eric De Jong, Richard Doyle, Tom Duxbury, Patricio Figueredo, Lou Friedman and his Planetary Society team, Mike Kaiserman, Rosaly Lopes, Sandy Miarecki, Scott Pace, Kevin Pope, Patrick Rayermann, Jan Smit, Bill Smythe, and Paul Weissman. Muchas gracias to my outstanding hosts and new friends from Argentina: Maria Isabel López and family, Analía Gomez Uria and César Maglia, Javier Fantini and family, Sergio Stinco and Susana Gimenez and family, Cumbre Radio, Jorge Calvo and his Proyecto Dino crew, and Bibi Canata. Teacher Louise Hopping and student-consultants David Bywalec, Brandon Kuc, Nicole Marshall, and Paige Stankus offered many excellent suggestions for improvement, as did content reviewers April Lynn Luehmann, Aileen Yingst, and Denton Ebel.—LJH

Cover photo: Planetary geologist Adriana Ocampo displays an impact spheroid found in the Albion Island rock pit in northern Belize. These spheroids tell a unique tale about what happened during the first few minutes after a space rock crashed into Earth.

Cover design: Michele de la Menardiere

Library of Congress Cataloging-in-Publication Data

Hopping, Lorraine Jean.
  Space rocks : the story of planetary geologist Adriana Ocampo / by Lorraine Jean Hopping.
      p. cm. — (Women's adventures in science)
  Includes bibliographical references and index.
  ISBN 0-531-16783-6 (lib. bdg.) ISBN 0-309-09555-7 (trade pbk.) ISBN 0-531-16958-8 (classroom pbk.)
  1. Ocampo, Adriana C. 2. Astrogeologists—Colombia—Biography—Juvenile literature. I. Title. II. Series.
  QB454.2.O23H67 2005
  559.9'092—dc22

                                                                    2005006644

Any opinions, findings, conclusions, or recommendations expressed in this volume are those of the author and do not necessarily reflect the views of the National Academy of Sciences or its affiliated institutions.

Printed in the United States of America.
1 2 3 4 5 6 7 8 9 10 R 14 13 12 11 10 09 08 07 06 05

ABOUT THE SERIES

The stories in the *Women's Adventures in Science* series are about real women and the scientific careers they pursue so passionately. Some of these women knew at a very young age that they wanted to become scientists. Others realized it much later. Some of the scientists described in this series had to overcome major personal or societal obstacles on the way to establishing their careers. Others followed a simpler and more congenial path. Despite their very different backgrounds and life stories, these remarkable women all share one important belief: the work they do is important and it can make the world a better place.

Unlike many other biography series, *Women's Adventures in Science* chronicles the lives of contemporary, working scientists. Each of the women profiled in the series participated in her book's creation by sharing important details about her life, providing personal photographs to help illustrate the story, making family, friends, and colleagues available for interviews, and explaining her scientific specialty in ways that will inform and engage young readers.

This series would not have been possible without the generous assistance of Sara Lee Schupf and the National Academy of Sciences, an individual and an organization united in the belief that the pursuit of science is crucial to our understanding of how the world works and in the recognition that women must play a central role in all areas of science. They hope that *Women's Adventures in Science* will entertain and enlighten readers with stories of intellectually curious girls who became determined and innovative scientists dedicated to the quest for new knowledge. They also hope the stories will inspire young people with talent and energy to consider similar pursuits. The challenges of a scientific career are great but the rewards can be even greater.

CONTENTS

Well Grounded in Planets

Adriana Ocampo has always dreamed of exploring planets. As a geologist, she's curious about their rocks and canyons and cliffs, but it's the impact craters that give her thrills—and chills.

Here on planet Earth, she helped find the Crater of Doom, a massive hole blasted out by a mountain-sized space rock 65 million years ago. This split-second collision led to a global disaster that killed off entire life forms, including the dinosaurs. Adriana is now searching Earth for the crater's ejecta—rock and debris from the blast that ended up all over our planet.

Meanwhile, she explores planets other than Earth through the electronic eyes of robotic spacecraft. As a teenager, she landed a job at NASA, the U.S. space agency. There she had a front-row seat when a Viking lander beamed back our first clear images from the surface of Mars. She later commanded the spacecraft Galileo to peer down at the icy crust of Europa, a moon that might harbor life!

All her science adventures began on a rooftop in Buenos Aires, Argentina, thousands of miles from NASA. How on earth did Adriana Ocampo make her dream of exploring space come true?

Adriana truly **believes**
that people will go to other planets,
and even to the stars,

and she **hopes** with all her heart
that she will be one of them.

LOS SUEÑOS (DREAMS)

BUENOS AIRES, ARGENTINA
LA TERAZZA LAUNCH SITE
MISSION: COLONIA LUNAR

driana Ocampo has waited a long time for this moment. All day, in fact. For hours she sat quietly in her white uniform, doing the necessary paperwork. Now it's finally time for some action. Her space mission is about to begin!

Adriana gulps down a *café con leche*, coffee with milk, and bread with her favorite treat, a caramel jam called *dulce de leche*. Then she quickly gathers a handful of tools from drawers and cabinets.

"*Vamos*, Tauro!" she calls to her copilot. Let's go!

With Tauro close behind, she hops up three flights of stairs to *la terazza*, the rooftop terrace where the launch pad is. Her spaceship is white except for the fuel tank, a round brown container of clear liquid that's taller than she is. As the commander, Adriana has full security clearance. She unhitches an iron guardrail and steps through the open door of the spaceship. Astronaut Tauro follows her inside. Astronaut Juanita, dressed in a shiny aluminum spacesuit, is already strapped into her seat.

Adriana Ocampo *(above)* wore white for her first communion, when Catholic children first take full part in a mass. On a family trip to central Argentina, she was fascinated by a water hole *(opposite)* where swirling water ground stones smooth against the rock wall.

Adriana sets down the tools, closes the hatch, and secures the red, black, and yellow clips that seal it. She makes a visual check of the control room. There's no window, not even a peephole, but she likes it that way. It helps keep her mind focused on the mission.

"Todos los sistemas listos," she announces, as she takes the commander's chair. All systems go.

The countdown begins almost immediately, in Spanish. *Tres, dos, uno . . . FUEGO!* LIFTOFF! Adriana bounces back and forth, up and down, as the capsule blasts into space. In seconds the three astronauts are on their way to the Moon!

~ *Anything Can Happen*

An oscilloscope measures electrical signals. (Oscillate means "to waver back and forth regularly.")

From past missions, Adriana knows that anything can happen in outer space. So she keeps a close eye on the oscilloscope [ah-SIL-uh-skope], a small but heavy metal box. It has a face full of dials and a green screen that glows. Adriana relies on the instrument to monitor activity outside the windowless space capsule. She peers at the glowing screen. A thin green line is oscillating—wavering up and down—on a grid. The wave pattern is stable, she notes, with no change in frequency. *Perfecto.*

She reaches over and adjusts a black dial. The wavy line starts jumping to a faster beat. Its peaks and valleys march quickly across the screen. There's something out there!

"Qué es esto?" Adriana wonders aloud. What is it? She picks up the voltmeter, a smaller monitoring device, and sets the dial at 0.000001—one-millionth! It's the most sensitive level possible. Then she touches one end of a wire to the red button and the other end to the black button. The needle jumps from left to right before hovering in the middle. Adriana knows that this can mean only one thing: *asteroide!*

2

She pictures the asteroid, a giant rock flying through space straight toward the capsule. If it hits, all three astronauts are doomed! Adriana warns Tauro and Juanita to hold on tight. Then she steers the capsule sharply off course, fighting to keep herself balanced.

She is just in time. The asteroid zooms harmlessly past the capsule and off into outer space.

~ Just as She Imagined

Back on course, the spaceship soon lands on the Moon. Adriana puts on her helmet and adjusts the communications antenna. Tauro's helmet has come loose, and Adriana orders him to be still: *"No te muevas, Tauro!"* She tightens the straps on his headgear.

"Listos," she says. We're ready.

Adriana releases the metal clips on the hatch, opens the iron guardrail, and steps out onto the Moon. It is just as she imagined: an eerie land of rocks and craters with mountains on the horizon. She surveys the lunar colony near the landing site. A big glass dome provides living quarters for all the colonists. Smaller domes include a laboratory for making oxygen, a barn for animals, and a greenhouse for growing vegetables. Lunar rovers are lined up neatly, ready to wheel across the rocky surface.

> She pictures the asteroid, a giant rock flying through space straight toward the capsule. If it hits, all three astronauts are doomed!

Today's mission, Adriana informs her crew, is to investigate a large crater that no human has ever visited. She and Tauro will drive to the rim and then climb down on foot to collect rocks and explore. Juanita will monitor the mission from the lunar colony.

Adriana retrieves her tools from the capsule, selects a rover, and heads out toward the crater. She hopes that one day she and Tauro will encounter more than rocks and craters. She daydreams about meeting intelligent life from other worlds.

She wonders what she would say to them. Or what *they* might say to *her!* She considers how they will communicate. They won't speak Spanish, that's for sure. Aliens would be more intelligent than humans, she reasons, since they know how to travel from star to star. And thinking is faster than speaking or writing. *Telepatía!* They'll use telepathy! Adriana imagines sending thoughts directly from mind to mind as she and Tauro step down into the crater.

Time passes quickly on the Moon. The Sun soon sets, and Adriana hurries to complete the mission and return to the spaceship. As she prepares for liftoff, a distant voice calls out.

"Adriana! Adrianita!"

~ *The Spell Is Broken*

The voice belongs to her *mami,* her mother. It's late evening, dinnertime in Argentina. Mami calls up the stairs again, and Tauro barks in reply. Adriana's copilot is her new puppy, a white fox terrier with brown and black spots. She removes her dog's little plastic helmet and tells him to be patient. Then she takes off her own helmet, a colander for straining water from pasta. The helmet's antenna is a meat skewer.

The spaceship, the lunar colony, the crater, the intelligent life forms—all of it melts from Adriana's mind. The ship transforms back into an old white sheet hung over an iron bar. The brown fuel tank becomes a water tank. The domed habitat, the oxygen lab, the barn, and the greenhouse return to kitchenware: upside-down bowls, painted jars and cans, saucers, and pot covers. The Moon rovers are toys made out of corks and wires. The tools are spatulas, tongs, and other utensils from Mami's kitchen.

The Moon mission is make-believe, of course, a typical play session for a girl with a very big imagination. Adrianita, which means "little Adriana" in Spanish, is about eight years old and plays in her rooftop lunar colony every day after school. It's the

Adriana (in the white hat) sits with her mother and sisters on the rooftop terrace where she built her Moon colony. The comic book she's holding is about a boy who has a nutty inventor for an uncle and dreams of being rich and famous.

mid-1960s, and the Ocampos live in a townhouse near Buenos Aires, Argentina.

Adriana picks up astronaut Juanita and tosses her into the spaceship. Juanita is a doll, a gift from Tío Ricardo, Adriana's uncle and godfather in Colombia. The doll has long, light brown hair and big eyes and came wearing a Scottish plaid outfit. Adriana can't stand dolls, so she made a little spacesuit out of aluminum foil, fashioned a tiny helmet out of a plastic bowl, and taped on another meat-skewer antenna. Juanita was no longer a doll. She was *una astronauta!*

Along with tools from Mami's kitchen, Adriana borrows supplies for her missions from Papi's workshop. Her father, Victor, is an electrician. He holds two jobs, sometimes three, often working from dawn until 11 P.M. In his workshop, he teaches people how to build and repair radios and televisions. The room is filled with so many interesting parts: wires, tubing, dials, magnets, antennae, voltmeters, glass vacuum tubes, and colored clips with "alligator teeth" for gripping. It doesn't take too much imagination to turn these gadgets into space equipment!

Adriana especially loves the oscilloscope. Papi uses it to measure electrical signals from outlets and machines. To Adriana it is the ultimate instrument for space exploration. With its dials and glowing green screen, it fits right into any science fiction adventure.

The Ocampos visited a tall ship called the *Sarmiento* when Adriana (standing in front of Mami) was about 8 years old.

~ We're Not Alone

Sometimes, Adrianita climbs up to the terrace just to look at the stars. She can pick out the Southern Cross, the most famous constellation in the Southern Hemisphere. She knows *Las Tres Marias*—the three stars of Orion's belt. But what really amazes her, night after night, is the extraordinary number of stars. Even in a big city, with millions of electric lights washing out the night sky, there are far too many stars to count.

There *has* to be intelligent life out there somewhere, Adriana believes. *"No estamos solos en el universo,"* Papi often assures her. We're not alone in the universe.

Some kids think that's weird, so Adriana keeps her space missions mostly to herself. Her older sister Sonia, who is 11, says there are too many problems on Earth to worry about outer space. She thinks it's better to help real people who are sick or poor than to dream about things that don't exist.

Three bright stars in a row—known as *Las Tres Marias* (the three Marias) to Adriana—make the constellation Orion easy to find. These stars form the belt of a hunter figure.

Adriana can't help it. She looks up at all those stars and has to ask, *"Quién vive allí?"* Who lives out there?

Adriana isn't the first person to ask that question. A lot of her ideas come from books. Every Saturday the Ocampos take turns reading something aloud—a poem, a chapter, or an original writing. On the terrace, when Adriana is by herself, she reads to Tauro. One of her favorite authors is Jules Verne. Mami gave her a book called *From the Earth to the Moon*. Verne wrote it more than a century ago, in 1865, yet the hero, Michel Arden, predicts, "Before twenty years are over, half the Earth will have visited the Moon!"

FROM THE EARTH TO THE MOON

DIRECT IN NINETY-SEVEN HOURS AND TWENTY MINUTES: AND A TRIP ROUND IT

BY
JULES VERNE

ILLUSTRATED

NEW YORK
CHARLES SCRIBNER'S SONS
1920

Projectile trains for the moon.

So far, in the mid-1960s, no one has set foot on the Moon. No one has even left Earth's orbit. Still, Adriana truly believes that people will go to the Moon, to other planets, and even to the stars. She hopes with all her heart that she will be one of them.

Mami and Papi always say, *"Hay que seguir los sueños."* You have to follow your dreams. And space exploration has been Adriana's dream for as long as she can remember.

The science fiction stories of Jules Verne have been translated from French into 148 languages! As a child, Adriana read them in Spanish. This illustrated English version of *From the Earth to the Moon* was printed in 1920.

~ The Centipede Dream

One night Sonia and Adrianita are tucked into their bunk beds, fast asleep. But Adriana is a restless sleeper. She is always tossing and turning, which annoys Sonia sometimes. Out of nowhere, an image pops into Adriana's sleeping head. In her mind she finds herself picturing a new kind of Moon rover!

Adriana wakes up with a start, jumps out of bed, and grabs a pencil and paper. She starts sketching, quickly, before the mental picture fades away. On the paper a centipede begins to take shape, or rather a vehicle that looks like that multilegged creature.

It has a series of glass spheres that are filled with breathable air and connected by a tube. Astronauts can walk from one bubble to the next through the tube. On the outside each bubble has two legs, one on either side, that fold up when the centipede isn't moving and come back down when it's time to go.

The rover can crawl slowly over the lunar surface, but Adrianita isn't finished. She has an extraordinary idea. Each sphere, she decides, can break off from the main part and zip around on its own. That way two astronauts can take off, explore a crater, and then zoom back to rejoin the centipede. *Fantástico!*

Adrianita thinks about how to build this incredible machine. In her mind she rifles through the kitchen drawers and cabinets. The tubing, she knows, she can find in Papi's workshop. Papi! She *has* to show him her centipede! She runs into the next room and taps on her father's shoulder. He's fast asleep. Adriana knows he has to get up at dawn to go to work, but she just can't wait that long.

Curious Scientists and Clever Engineers

When Adriana designed, built, and tested a rover, she was engineering. When she explored a Moon crater in her mind, she was imagining herself as a scientist.

What's the difference between an engineer and a scientist? Here's one way to think about it: An engineer designs spaceships that can take human beings to the Moon. A scientist wants to know what the human beings will find when they get there.

Science comes from the Latin word for "knowledge." So what do scientists want to know? Everything. They investigate how the universe works. They ask questions and seek answers by experimenting, observing, and gathering data. Science is about being curious.

The word *engineer* is related to the word *ingenious,* which describes someone with natural cleverness or skills. Engineers use science and math knowledge to invent, design, build, test, and improve things.

Like what sorts of things? Look around. Engineering involves any object not found in nature, such as buildings, roads, bridges, chemicals, and every kind of machine.

"Papi!" she says.

He rolls over and sleepily mumbles, "Anita?" Anita is Papi's nickname for her.

Adriana proudly shows him her drawing.

Papi glances at it and tells her it's very good. He promises to take a closer look in the morning. *How can such a small child be thinking these things?* he wonders before falling back to sleep.

Adriana can't wait to build and test her rover. How will

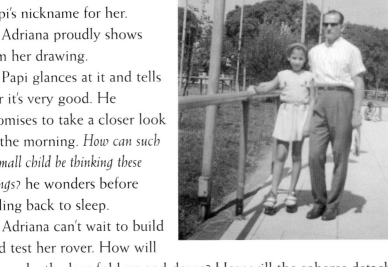

Papi worked three jobs—as a teacher, an electrician, and a salesman—but he still found time to teach Adriana to rollerskate. He also bought her a chemistry set and a little telescope with a carrying case. Adriana treasured both gifts.

she make the legs fold up and down? How will the spheres detach and reattach? Her mind never stops wondering, tinkering, and imagining—often all at the same time. Many of her earliest thoughts, as a very young child, led to both discovery and danger. One idea was downright deadly.

Adriana was an **optimist,**

someone who always
looks on the **bright** *side.*

HERE'S A DANGEROUS THOUGHT...

Adriana Cristian Ocampo was born on January 5, 1955, in the port city of Barranquilla, Colombia, a tropical country in South America. Less than a year later, the family moved south to Argentina, her mother's home country.

One day, when Adrianita was four or five years old, she heard Mami complain about the flies. That gave her an idea. Mami was always telling her and Sonia to never eat the calla lilies: "*Nunca comas las calas!*" They can kill you, Mami warned.

Adrianita thought, *If those flowers can kill people, maybe they could kill the flies.* She would make a *calas* bug spray for Mami! In the garden she recognized the flowers by their white, funnel-shaped petals with an orange knob in the middle. For a killer plant, *calas* looked beautiful. She picked a handful and brought them to the patio. Then she looked in cupboards, drawers, and closets until she found an empty plastic bottle. The cap had a little trigger: Press it, she knew, and the liquid would spray in a mist. *Perfecto!*

Adrianita removed the cap and broke open the flower stems with her fingers. Drops of white liquid—*el veneno!* the venom!—dripped into the bottle. There wasn't much of the liquid poison. So Adriana added water and, for good measure, a splash of rubbing alcohol. *Está listo!* That should do it.

When Adriana was still a toddler (holding Mami's hand, *above*), the Ocampos lived with Adriana's grandparents on her mother's side. Adriana had a "strong mind, a strong will, and a strong personality," says her mother (pictured with an older Adriana, *opposite*).

She took the bottle outside and waited for a fly to land. When one did, she aimed the sprayer and pulled the trigger. *En la marca!* Bull's-eye! A cloud of mist drenched the fly. The insect flopped around and then stopped moving. Her bug spray worked! *"Fantástico!"* she said.

When Mami walked outside to see what was going on, Adriana proudly presented her bug spray and explained how she made it. Mami took the bottle and said it was very dangerous. *"Es muy peligroso!"* she scolded. Then she made a new rule: *"No toques las calas!"* Don't even *touch* the calla lilies!

Calla lilies are beautiful but deadly. Their milky juice contains crystals that can severely burn and swell the skin, mouth, lips, and throat on contact. If eaten, any part of the plant can cause sharp stomach pain at best and death at worst.

~ Chicos and Chicas, Boys and Girls

Adriana loved to play by herself, but she also made friends easily. She found that most *chicas*—girls—had no interest in space adventures or playing with big, ugly rhinoceros beetles (her favorite animal). That included Sonia and their new baby sister, Claudia. So Adriana usually played with *los chicos,* the boys.

Two exceptions were a pair of girls a few years older than Adriana. She entertained them with stories about magic carpet rides and space adventures. One day she asked what they wanted to do when they grew up. They didn't know. Adriana asked again. Surely they liked something. Mami and Papi were always saying that everyone has a mission in life.

One girl finally said she might become *una enfermera,* a nurse. *"Porqué no una doctora?"* Adriana countered. Why not a doctor?

It was true that in the 1960s doctors were usually men and nurses were usually women, but the Ocampo girls were taught they could be anything they wanted. There were no excuses for not trying to succeed, her parents said, and they had plenty of examples to prove it: There was Domingo Sarmiento, a poor peasant who became president of Argentina. Thomas Edison was terrible at school and lost nearly all of his hearing, yet he was a brilliant inventor. As devout Catholics, her parents often pointed to Jesus Christ as an example of success—a simple carpenter who was now worshiped around the world.

Papi believed that for some people, life was like riding an elevator. Closed inside the elevator, they zoomed straight to the top without seeing anything on the way up. Other people took the escalator of life, he told his three daughters. They rose floor by floor, working extra hard at every stage, but they experienced everything life had to offer. When they reached the top, they were much richer for the ride.

While serving in the Colombian Navy, Victor Ocampo was stationed in Buenos Aires, Argentina, where he met Teresa Uria, a teacher. The couple married *(above)* and lived in Colombia for several years, then moved to Argentina just after Adriana was born.

~ School Rules

Adriana understood what it meant to work hard at every stage, especially when it came to school. Every morning the classes lined up to raise the Argentine flag and sing the national anthem. The children stood in rows, with the best students in front. Sonia, at the top of her class, was the *abandera*, the flag bearer. Adriana struggled to make average grades.

She was smart, but she was a bit of a rebel. She didn't like being told what to do, and Argentine schools were very strict. Every student wore a *guardapolvo*, a white robe over their clothes. When the teacher entered, the class stood and said, *"Buenos días, profesora."* Good morning, teacher. Students had to stand up to answer a question and no one talked out of turn. Adriana learned that rule the hard way—by getting her hand slapped with a ruler. Even recess felt like prison to Adriana. A *celador*, a hall monitor, knew every boy and girl and reported every misdeed.

Adriana felt like breaking all the rules, but that was only part of the problem. She had a rich imagination. To her the lessons were

All the girls in Adriana's junior high class wore their *guardapolvos* (white robes) for the school photo. Adriana is in the second row from the top, fourth from the right.

painfully boring, so she daydreamed to escape them. At home, learning was far more fun. Her mother, Teresa, had taught in a Montessori school before becoming a full-time homemaker. Montessori is a style of education that encourages children to think and imagine and discover for themselves. Teachers don't stand in front of the class and talk; instead, students choose a project in math, science, or another area and work on it as long as they like.

Teresa encouraged her three daughters to make or do something new every day. Together they wrote songs and poems, listened to opera music, crafted goofy creatures out of potatoes and toothpicks, and floated boats made of corks.

Adriana learned to repair radios and hook up electrical circuits in her father's workshop. She helped Tío Santiago, her uncle, fix his 1955 Oldsmobile sedan. Her parents didn't own a car, so it was Tío Santiago who gave Adriana her first automobile ride. She loved it! In fact, she loved any kind of vehicle. She drew designs for flying machines, including winged space shuttles that could land on the Moon and return. (Real space shuttles wouldn't be invented for another decade!)

One day Adriana decided to test paper airplanes, another experiment that turned out to be *muy peligroso*—very dangerous. She and Tauro climbed to the highest point of the roof, above the make-believe lunar colony, to launch the planes. Curiosity had overruled any fear of falling, but luckily a neighbor spotted her and called her mother. Under Teresa's watchful eye, both Adriana and Tauro climbed down safely.

What a contrast to school! There, Adriana was told what to study, when to study it, and how to study it. Her struggle with formal education would become a lifelong challenge.

On a typical Saturday in Buenos Aires, Adriana (on the left) and her sister Sonia went horseback riding in the morning and visited the futuristic-looking plantetarium *(top)* after lunch. During the night-sky show, Adriana sat back and dreamed of traveling to the stars.

~ A Risky Move

In the late 1960s the economy of Argentina began to weaken, and the government became unstable. Teresa and Victor worried about tough times ahead, so they took a risky step: They decided Victor would move thousands of miles away, to the United States, to look for a better job. If he found one, Teresa and the three girls would follow. He moved to Los Angeles, California, in September 1968, when Adriana was 13 years old.

Victor and Teresa sometimes talked for a few minutes by telephone, but international calls were very expensive. Instead the Ocampos wrote long letters. Adriana's letters were full of technical questions, such as how to repair the washing machine. As a little girl, she had asked her father countless times, *"Cómo funcióna?"*

At Vicente López junior high, Maria Isabel *(left)* wanted to fly jets while Adriana *(right)* dreamed of becoming an astronaut or engineer. All three careers were off-limits to Argentine girls in the late 1960s.

How does it work? Now, barely a teenager, she had to know, *"Cómo lo reparo?"* How do I repair it?

While her father was away, Adriana attended Colegio Nacional de Vicente López, a four-story, brown brick junior high school. It happened to be at the corner of Tomás Edison Street, the Spanish name for Papi's favorite inventor.

While many of her classmates worried about fitting in, Adriana went out of her way to be different. Her best friend Maria Isabel and other girls had long hair, but Adriana cut hers short. She wore a funky cap and trendy, knee-length Bermuda shorts. She even bought a pink bikini to wear on family vacations.

~ "A Very Evolved Mind"

In *primero año* (seventh grade), Adriana studied science, math, geography, history, government, handwriting, composition, literature, French, and Latin. She had to pass every class to move up to the next grade, *segundo año*. She didn't. In December 1968 she learned she had failed not one, but two subjects: geography and history.

The teacher told Teresa, "Your daughter has to prepare more for school and work harder. She's very slow."

Both Victor and Teresa knew better. Their middle daughter had what Victor liked to call "a very evolved mind." Teresa understood that Adriana liked to do things her own way. "She has a strong mind, a strong will, and a strong personality," she explained.

As Adriana turned 14, she used both her "very evolved mind" and "strong will" over the summer to study for makeup tests. She had gone through school with the same group of students each year and couldn't imagine staying behind with another class. To her great relief, she passed both tests and entered *segundo año* in March, the beginning of the school year in Argentina.

All the students in *segundo año* had to take a test to find out what skills and subjects they were good at. The results determined which high school they could attend. Sonia, with her perfect grades, was in a high school for future lawyers and doctors. Adriana's results showed that she was good at numbers and that her mind was analytical; she could break down a big problem into steps and solve it. That's an excellent skill for an engineer, and she longed to go to the technical high school. There she could learn how to design machines and buildings—something she loved to do for fun. Unfortunately, engineering was for boys only.

> She decided to do something about it. She wrote a letter to NASA, the U.S. space agency, and in Spanish asked if she could work there, ideally as an astronaut.

Adriana was placed in a school for *perito mercantil*—business studies. She learned to keep track of money—the *pesos* in and *pesos* out of a company—which required calculating by hand, over and over. For a girl who dreamed of rocketing into space, it was pure torture. Adriana thought the classes were mind-numbing!

She decided to do something about it. She wrote a letter to NASA, the U.S. space agency, and in Spanish asked if she could work there, ideally as an astronaut. It didn't matter to her that every astronaut spoke either English or Russian, not Spanish. She didn't care that they were all grown men and she was a petite teenaged girl. So what? She knew she could be an astronaut!

Weeks passed. There was no reply, but Adriana continued to follow news of the space missions closely.

~ Men on the Moon

Adriana watched on TV as American astronauts Buzz Aldrin (below) and Neil Armstrong collected rocks and set up experiments on the lunar surface. Next to Aldrin in the photo is a solar-powered seismograph for sensing moonquakes.

On July 20, 1969, two American astronauts, Neil Armstrong and Buzz Aldrin, landed on the Moon! The Ocampos were the only family in the neighborhood with a TV, so Adriana invited friends over to watch the big event. The old black-and-white set had a tiny screen and "rabbit ears"—two moveable antennae—but a television was a luxury few people could afford.

Close to midnight in Buenos Aires, Adriana watched, eyes big and round, as Neil Armstrong stepped out of the lunar lander in his bulky white spacesuit and backed down the ladder. When he

got to the last rung, he hopped off and landed with a bounce. It was the first step by a human being on another world.

Adriana simply said, *"Fenómeno!"* Extraordinary! It was her favorite word. Long after her mother and sisters went to bed, she stayed up to watch the astronauts explore. There were rocks and craters, just as she had imagined, and some surprises, too. The sky was inky black! The Sun was so glaring and bright! There was dust *everywhere.*

An *Apollo 11* camera *(foreground)* snapped the first close-ups of a lunar crater. The Moon landings proved that the holes were made by crashing asteroids and comets, not by volcanoes.

Adriana was more certain than ever that she wanted to be a space explorer. In the coming weeks, her dream began to slowly come true, first in the form of a package and then with a phone call.

The package—all the way from the United States!—arrived addressed to "Miss Adriana Ocampo." It was from NASA! Adriana tore open the envelope and found a letter in English. She couldn't read it, but she dug further and found the real prize: a special postage stamp. It showed an American astronaut, Neil Armstrong, making that first giant step onto the Moon.

FIRST MAN ON THE MOON — UNITED STATES

In 1969, at age 14, Adriana was thrilled to receive a special stamp from NASA in honor of the first Moon landing.

Then in December, Teresa got a phone call from Victor. He had a job at a car factory and was finally ready for his family to move to California. There was no time for discussion. The phone call was too expensive. So just like that, in two minutes, Adriana's whole life changed. She would soon be living in the country that had sent astronauts to the Moon!

~ Her First True Adventure

Adriana was about to have her first *real* adventure, though not everyone saw it that way. Sonia had just graduated from high school, and she followed world news closely. In the 1960s, Los Angeles and other American cities experienced violent riots. Angry people smashed storefronts, started fires, and attacked each other in the streets. That really worried Sonia.

Tío Santiago was worried, too, for his three nieces. He said that he didn't like the hippies, those young people wearing wild clothes and taking drugs and dropping out of school.

Victor wrote to his family that people can get the wrong idea from the news and that Los Angeles would be a good home. Like her father, Adriana was an optimist, someone who always looks on the bright side.

> So just like that, in two minutes, Adriana's whole life changed. She would soon be living in the country that had sent astronauts to the Moon!

Sonia thought Adriana and Papi were dreamers. Her view, she firmly believed, was more realistic.

So who was right? Was the United States violent and dangerous? Or was it the land of adventure and opportunity? The Ocampo women would have to see when they got there. Teresa and her daughters each packed one suitcase to take on the two-day flight from Buenos Aires to Los Angeles.

Adriana wondered what to pack—and what to leave behind forever. She needed clothes, of course. She added her Jules Verne books and a few other favorites to the suitcase. Her mother had encouraged her to keep a notebook full of her thoughts and ideas and poems. In the notebook, she also collected phrases that she liked to share with others: "Work hard." "Have a good attitude." "Face your problems." Another page titled *"Opcióñes"* was a list of options—things Adriana hoped to do with her life.

She carefully placed the notebook in her suitcase. Those sunny thoughts and sky-high dreams were about to take Adriana Ocampo further than even she could imagine.

Apollo 17 *was ready for* **launch**
at the **Kennedy Space Center** *in Florida—*

and Adriana Ocampo
would be there to **watch** *it!*

WELCOME TO NASA

3

In December of 1969, Teresa Ocampo and her three daughters—Sonia, Adriana, and Claudia—stepped off the plane at Los Angeles International Airport. Adriana kept her eyes and ears open to soak in every moment of this adventure. The strange sounds of English came at her from all directions, and she felt lost. But then a border guard smiled and said, "Welcome to the United States!" She understood *those* words. They were proof that she really was in America.

Victor Ocampo had waited at the airport all day, even though he knew his family wasn't due to arrive until the afternoon. As soon as Adriana spotted Papi, she charged him like a bull. The Ocampos couldn't stop hugging, kissing, and crying. No one could speak. They had been separated for more than a year.

Victor led his family out of the airport to the only car he had ever owned: a 1968 Volkswagen Beetle. With its hump-shaped body and bright red color, the Beetle looked like an oversized ladybug. Somehow, a father, mother, three daughters, and all their suitcases fit inside the tiny two-door car.

As Victor pulled out onto the Los Angeles freeway, Adriana spoke for the first time. She asked, *"Dónde está NASA, Papi?"* Where's NASA?

Victor smiled. His space-crazy daughter hadn't changed too much.

Adriana *(front and center, above)* and her fellow Space Explorers toured the Kennedy Space Center in Florida before watching the night launch of *Apollo 17 (opposite)*. The tiny, cone-shaped Apollo capsule, barely big enough to hold three astronauts, sits atop a Saturn V rocket as tall as a skyscraper.

~ Seeing Black and White

The family rented a small apartment in East L.A., where many other Spanish-speaking immigrants lived. Most of their neighbors were from Mexico and Central America, and Sonia was surprised that they didn't like to speak Spanish, their native language. They were proud that they could speak English and looked down on the Ocampo girls, who were still learning the language.

Though she couldn't understand the teachers, Adriana attended the ninth grade at Cudahy Junior High School. In the cafeteria, she was amazed to see black-skinned African Americans, brown-skinned Hispanics, and fair-skinned whites sitting at separate tables, like sorted socks. She didn't have a clue why. The whole idea of different races was new to her. In Argentina she had known only one student with dark skin—a *mestizo*, a boy with mixed Spanish and Indian blood. Almost all Argentines are of European descent—Italian, German, Spanish, and British.

Papi liked to say, "You can see all the faces of the world in the United States." Now Adriana knew what he meant.

Other things were different, too. On one of her first days in class, she stood up to answer a question from the teacher. Everyone laughed. In Argentina, Adriana and her girlfriends had walked hand in hand or arm in arm. Here, she learned quickly, that just wasn't done.

For the most part, Adriana shrugged off all these differences, but one day a student began waving the sharp points of a compass—the kind you draw circles with—like a weapon! Just weeks ago in Argentina, talking out of turn had been a serious offense. This kid was threatening to stab people! He didn't,

In 1970, soon after moving to Los Angeles, Adriana's second must-see site was Disneyland. (The first question she asked Papi was, "Where's NASA?") She's in pink, standing next to Sonia (in red), Claudia (in blue), and Teresa.

24

luckily, but the shock stayed with Adriana for a long time.

~ Rosy Pasadena

Victor and Teresa decided to move to South Pasadena, a city just north of Los Angeles. The streets were lined with tidy, one-story houses, green lawns, pretty gardens, and fences. Claudia went to the local elementary school, and Sonia entered UCLA—the University of California at Los Angeles.

Adriana started tenth grade at South Pasadena High School. The large, modern high school had tennis courts, sports fields, an auditorium, and well-equipped classrooms with science labs and workshops. It was also right next door to the Ocampos' new church, Holy Family Catholic Church.

Adriana signed up for English as a Second Language classes with the other foreign students. Shop class (woodworking), drafting (drawing plans), and auto mechanics were for boys only, but she asked for special permission to take those classes. Though she was the only girl and spoke broken English, she loved building and fixing things and felt at ease around the boys. It didn't matter how different she was—or they were.

With other teenaged friends, Adriana helped decorate a float for Pasadena's famous Tournament of Roses Parade, held every January 1. The millions of fresh-picked roses were stunning, but it was another local attraction that *really* drew her eye. Pasadena is home to the California Institute of Technology (Caltech), the university that runs the Jet Propulsion

Adriana (on the right in the top photo) loved the flowering tree in her front yard in South Pasadena. In the bottom photo she's creating an oil painting of the Sun.

Laboratory (JPL), a NASA center. Just by living in the area, Adriana was a step closer to her dream of exploring space!

One day some Space Explorers even showed up at her high school. The Boy Scouts of America and JPL had just created Space Exploration Post 509 and were recruiting 14- to 20-year-old members. Unlike the technical school in Argentina and the boys-only classes at South Pasadena, Adriana was happy to discover that girls were allowed—and even welcomed.

~ Cómo Funcióna?
How Does It Work?

The Jet Propulsion Laboratory is the size of a small town, with more than 4,000 workers, three cafeterias, a bank, and a medical center.

Papi drove Adriana to a meeting at JPL, a large complex of buildings spread over a 177-acre campus. The NASA center was nestled in the desert foothills of the San Gabriel Mountains, about 20 minutes from Pasadena. It was almost all off-limits to the public, and the meeting was confined to the von Kármán Auditorium, named after one of JPL's founders.

The Jet Propulsion Lab Is Born

In the early 1940s, during World War II, Theodore von Kármán *(right)* and other Caltech engineers founded a lab to design rocket missiles. The missiles had jet engines, and anything that supplies the force to push an object is a propulsion system, so the facility was named the Jet Propulsion Lab (JPL). The name stuck,

but as the Space Age dawned, JPL switched its mission from propulsion systems (the engines that make rockets go) to payloads—the things rockets carry, such as satellites and robotic spacecraft. The first successful U.S. satellite, *Explorer 1,* soared into Earth orbit on January 31, 1958, eight months before NASA was founded.

Almost as soon as Adriana walked inside, she stopped in her tracks. She couldn't believe what was right in front of her eyes. She focused all her thoughts on the strange machine. It looked like an oversized ceiling fan with instruments clumped in the middle. "Satellite," the sign said. *Un satelite!* Adriana had seen them in pictures, but now she was standing right next to a real one!

Without saying a word, she walked all around it, staring from every angle. She touched one of the four winglike solar panels. They captured the Sun's energy, she knew, and turned it into electricity. *But where are the batteries to store the energy? And the propulsion system?* Something had to push the machine forward in space. *That boxy object on top must be a motor or engine,* she figured. She looked for thrusters—gas jets for steering—and spotted the little nozzles at the tips of the solar panels.

Adriana recognized the radio antenna—a white dish-shaped object—and another antenna that looked more like the meat skewer on her homemade helmet back in Argentina. There were several weird cameras and other instruments that she had never seen before.

In a whisper, she asked Papi that familiar question, *"Cómo funcióna?"* How does it work? Papi just smiled. Adriana was about to find out for herself. In the spring of 1972 she began attending the weekly, Thursday-evening meetings of Space Exploration Post 509 in the von Kármán Auditorium.

~ *Laboratory Animals*

After the very first meeting, Adriana told her mother, "This is what I want to do!" "This" meant aerospace engineering, which was the main activity of Post 509. During the next few years, she would be firing model rockets, programming robots, and building circuit boards. With their Post 509 ID badges, the teenagers also got to tour the "off-limits" parts of the massive NASA center.

After passing through the security door, Adriana stepped out onto a central lawn, several blocks long, that was ringed with neatly trimmed trees and shrubs. The structures surrounding the lawn were all business—a gray and white jungle of high-rise offices, high-tech labs, NASA trailers, and roads named after space missions (Mariner, Explorer, Surveyor, and Pioneer).

Even so, Adriana often caught glimpses of wildlife: lizards, snakes, raccoons, and opossums. Families of mule deer—small deer with comically large ears—lived on the central lawn. A voice over the loudspeaker occasionally warned of a cougar or coyote sighting. Adriana thought, *If wild animals are happy to be at JPL, it must be a great place for humans to work, too.*

Adriana soon understood why JPL was so huge: Behind every space mission stood hundreds of people. In May the Space Explorers toured the Space Flight Operations Facility (SFOF) in Building 264. The SFOF is a mission control center for robotic spacecraft, bundles of instruments that zoom across the solar system and transmit data and images back to Earth.

> After the very first meeting, Adriana told her mother, "This is what I want to do!" "This" meant aerospace engineering, which was the main activity of Post 509.

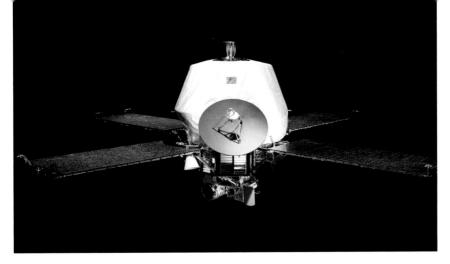

At JPL, Adriana learned how robotic spacecraft like *Mariner 9 (left)* are designed to explore the solar system. They navigate by the stars, gather power from the Sun through winglike solar panels, steer by way of gas jets, and "phone home" through radio dishes and other antennae. At a planet or moon, their instruments gather data until the power runs out.

Adriana learned that *Pioneer 10* was heading to Jupiter and would be the first Earth-made object to pass through the asteroid belt. No one knew if it would slip through the band of space rocks safely. (It would.) *Mariner 9*, now in orbit around Mars, was sending back the clearest images of the red planet ever seen. An engineer on the mission, Mike Kaiserman, soon became the head advisor of Post 509.

~ The Human Side of Space

Many of JPL's robotic missions were spectacular successes, but they weren't in the public spotlight. Human spaceflights, like the *Apollo 11* Moon landing, gripped the world's attention. That feat had been followed by four more Moon landings and one near-fatal accident. Now, in December 1972, the last human mission to the Moon was about to begin: *Apollo 17* was ready for launch at the Kennedy Space Center in Florida—and Adriana Ocampo would be there to watch it!

With car washes and raffles, Post 509 had raised money for the cross-country trip. Mike Kaiserman and other advisors drummed up donations. Over the summer Mike had also started a work program for teenagers. Adriana and other Space Explorers earned a little better than minimum wage working at JPL.

On the way to Florida, the group stopped in Houston, Texas, to tour the Manned Spacecraft Center, the NASA center where astronauts train and mission controllers manage human spaceflights. (It would be renamed the Johnson Space Center the following year, 1973.) Adriana watched, fascinated, as technicians ran tests on the

At the Kennedy Space Center, the Post 509 members were all "floating on a cloud," noticed advisor Mike Kaiserman (kneeling, in the pink shirt). "Their minds were wide open, and they couldn't learn fast enough." Adriana is to the left behind him, wearing a yellow sweater.

Apollo 13 capsule to see what went wrong. In 1970, less than two years before, its three astronauts had barely made it back to Earth after an explosion. The accident gave rise to that familiar line, "Houston, we've got a problem."

Finally, on the evening of December 6, Adriana found herself sitting a few miles from Launch Pad 39A at the Kennedy Space Center. The Space Explorers were among 700,000 observers who had flocked to Florida to witness the *Apollo 17* launch. People lined the roads, beaches, fields, and parking lots with their tents and pickup trucks and trailers. The Space Explorers had bleacher seats along the Banana River.

~ Goodbye, Apollo

Besides being the last Moon mission, *Apollo 17* was the only mission that included a scientist—a Caltech geologist named Jack Schmitt. From her rocket projects, Adriana understood that he and his two crewmates were sitting on a bomb. That is, they were inside a cramped, cone-shaped capsule at the tip of a Saturn V, NASA's biggest and most powerful rocket. To propel the tiny capsule into orbit, the Saturn V had to explode like a bomb, with Earth-rattling force.

Apollo 17 was also the only night launch, and the controlled explosion promised to be stunning. The countdown began a couple hours before liftoff and ticked away over the course of the evening. Then at 9:53 P.M.—just 60 seconds to launch!—it stopped. The mighty Saturn V engines shut down. Adriana didn't know why, but she did know that it was unusual. This had never happened so close to liftoff.

Two fidgety hours passed as technicians worked on the problem—a computer glitch, as it turned out. The Space Explorers had traveled all the way from California to see this launch. Now they worried that the mission would be scrapped.

Finally at 12:33 A.M., the Saturn V blasted off. *It lit up like a torch,* Adriana thought. In an instant, the night became as bright as Florida sunshine, and Adriana could feel her whole body vibrate. She felt pressure on her chest from the shock wave as the rocket powered its way upward, off the launch pad. Its bright orange tail, half a mile long, soared into the night sky.

After a blazing night launch, *Apollo 17* sped to the Moon in five days. The landing was successful, but NASA canceled future Moon missions due to high costs and fading public interest. Since December 1972 and as of this writing (2005), no human being has left Earth orbit.

The astronauts in *Apollo 17* were on their way to the Moon, and Adriana was floating on a cloud! She knew she would remember this magic moment—a dream come true—for the rest of her life. The Apollo era may have been ending, but her adventures as a Space Explorer were just beginning. She etched the exact starting date in her memory: June 25, 1973.

At age 18, Adriana found herself in the perfect place at the perfect time

with a useful background in electronics and computers.

A Genuine Engineer

As the Apollo missions faded into the history books, JPL's robotic spacecraft zoomed into the spotlight. By the early 1970s, the NASA center had mushroomed to 4,000 employees. Yet of the top 1,405 people, only 14 were women and 68 were Latinos (Latin Americans), like Adriana, or other minorities. The push for equal rights was at a peak, and JPL was seeking to hire a more diverse pool of talent. Also, in the summer of 1973, Mike Kaiserman's work program for Post 509 teenagers was expanded.

At age 18, Adriana found herself in the perfect place at the perfect time with a useful background in electronics and computers. Right after she graduated from South Pasadena High School, she landed a permanent job at JPL, starting on June 25, 1973.

She was still under the umbrella of Post 509, but she had a photo ID, a desk, a chair, and a title. She began as a technical aide on a radio astronomy project called Very Long Baseline Interferometry (VLBI). The idea was simple: the bigger the ears, the better the hearing. Giant dishes all over the planet collect the radio waves that some stars give off naturally. VLBI linked lots of dishes by radio to a single dish in orbit. The system acted like one Earth-sized "ear" for listening to the stars.

While learning English, Adriana had to work extra hard to graduate from South Pasadena High School in 1973 (opposite). Victor and Teresa bought Adriana a small used car for her 18th birthday, and Adriana taught her mother how to drive.

As a teenager among adults with Ph.D.s, the highest college degrees, Adriana's role was a small one: She helped collect and track the incoming radio waves. She was learning a lot, but to advance at JPL, she needed a college degree. She enrolled part-time at Pasadena City College, a two-year community college.

~ Motorcycle Maintenance

In Pasadena, Adriana kept up the horseback riding skills she had learned in Argentina at a sports club paid for by the government.

To cruise between college classes, her JPL job, Post 509 activities, and home, Adriana drove a black Karmann Ghia. The small European car was a great gift from her parents, but it broke down far too often. Adriana typically looked at the bright side.

"That car is teaching me a lot about auto mechanics," she told friends, without a hint of complaint.

Still, she needed backup transportation. Through a classified ad in *The Universe*, the JPL newspaper, Adriana bought a motorcycle for $50. At that price she wasn't expecting a deluxe model, but what she got was a frame and a box. The box contained the motor and assorted parts, which Adriana painstakingly pieced together.

Motorcycle maintenance, classes, and radio signals kept her very busy, but every once in a while she had a rare evening free. Adriana and Sonia attended a few social events at Caltech together. Men far outnumbered women at the technical university, and the South American sisters drew a lot of attention. Adriana was attracted to men who could talk intelligently, so she was in the right place. She dated a couple of brainy guys—a chemist and a mathematician who liked to play hockey—but neither one seriously.

Meanwhile she had been making lifelong friends through Post 509. A boy named Patrick Rayermann had an astonishing memory for dates and names. A newcomer named Scott Pace looked up to Adriana as a senior member of the club.

34

They were all working on Post 509's first real engineering project: a weather station. Of course, Adriana had quickly learned that nothing at NASA has such a simple, straightforward name. The projects, the buildings, the job titles, and the equipment all had codelike acronyms. The name NASA itself stands for the National Aeronautics and Space Administration.

The weather station was officially named the Environmental Monitoring Complex (EMC). The EMC filled a large trailer parked in the corner of the JPL visitors lot. Pat Rayermann thought its NASA-gray color looked more like "sickly pink," but the Space Explorers were excited to have their own office and workshop.

It was a "complex" for good reason. The project had these seven committees: facilities (the trailer), funds and supplies, sensors (devices to collect and transmit data), electronics, programming, applications (finding uses for the data), and documentation (reports). The positions on the committees had filled quickly. However, in September 1973, the teenaged leaders of the EMC project—the coordinator and vice coordinator—both quit. Someone had to give orders, oversee the committees, and make sure the project got done on time and on budget.

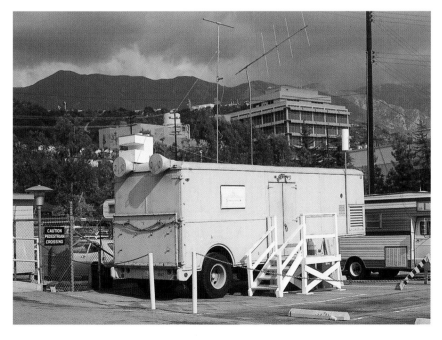

The outside of the EMC trailer looked drab, but Adriana and her Post 509 pal Pat Rayermann painted the inside bright yellow.

~ Who, Me? A Leader?

Head advisor Mike Kaiserman *(below)* convinced Adriana she had the right stuff to be a leader. In turn, Adriana convinced other teenagers to join her at Post 509 *(bottom photo)* for space projects involving real satellites and other NASA technology.

During a Post 509 meeting at the von Kármán Auditorium, advisor Mike Kaiserman spotted Adriana sitting quietly in the back. He had gotten to know her better on the *Apollo 17* trip the December before. She struck him as a shy, attractive girl with a sweet smile who was mature for her age. Mike also noted that she "seemed very in tune with the idea of teamwork." She blended in with groups, despite coming from a different culture.

Mike thought a leadership role on the EMC project would help bring out Adriana's skills. He whispered the idea in her ear.

Adriana balked. She didn't think of herself as a leader.

"I think someone else might be better," she whispered back.

Mike pointed out that if it didn't work out, she could switch to another job. "There's no harm in trying."

Adriana knew he was right. She had nothing to lose and a lot of experience to gain. She nervously agreed.

The EMC had three goals, each more challenging than the last. The first part was an automatic weather station for measuring temperature, wind speed, and so on. In NASA-like manner, it had an acronym. It was called the ARMS, which stood for Automatic Real-time Meteorological Subsystem. The ARMS was well under way when Adriana took charge. Part two was still on the drawing board: The APTRSS, or Automatic Picture Transmission Receiving Subsystem, would display weather images from satellites at the same time the pictures were being made. Part three, the ISMS, would measure the city's infamous air pollution. Its letters stand for Instrumented Smog Monitoring Subsystem.

An EMC team installs weather sensors on a giant radio antenna behind JPL. The weather data—temperature, wind speed, and so on—was transmitted automatically to the EMC trailer.

If all went well, sensors would collect weather and smog data on a mesa, a flat-topped mountain behind JPL. A radio antenna would transmit the data to the EMC trailer. A computer would combine, process, and store the local data and the satellite data for other computers to access. (Personal computers wouldn't be invented until the end of the decade, so the team had to build a computer from scratch.)

Adriana and her team presented their plan to William Pickering himself, the director of JPL. They convinced Dr. Pickering that they could meet all three lofty goals. The final review would be in six months. If the teenagers' work didn't pass review, all funding would be cut and the EMC project killed. That was the rule for all JPL projects, including the multimillion-dollar spacecraft missions.

Inside the EMC trailer, Pat Rayermann listens to incoming radio signals on the Automatic Real-time Meteorological Subsystem (ARMS). The ARMS includes an oscilloscope, which displays the wave pattern of the signals on a small screen.

~ User-Friendly Coaxing

In November during a geology field trip to Arizona, Adriana asked Pat Rayermann to be her second-in-command. He was already Post Historian and Post Recording Secretary, but he happily accepted.

Back at JPL, Adriana and Pat worked well together. Pat admired his partner's ability to "help people feel good about themselves and their contributions." He loved that Adriana

could give orders without making people feel like they were being pushed. He called it "user-friendly coaxing."

Scott Pace also looked up to Adriana. He knew that a lot of the teenaged engineers "didn't have the best social skills." Like Pat, he was amazed when Adriana would look at someone and say, "It would really be good if you did this," and the person would do it!

Adriana's sunny attitude helped. Her soft Spanish accent did, too; it had a calming effect, which came in handy when opinionated members argued. She also clearly knew what she was doing. People trusted her judgment and respected her decisions.

With a tight budget, a tall order, and little time, everyone's job was critical to success. Joe Coulombe and Greg Stark loved to build electronic circuits in Joe's garage. Some of their circuits were programmable computer "brains." Doreen Lane used other circuits to make weather sensors. Her rain gauge, dust collector, and barometer automatically fed data to the radio antenna. She was working on the wind, temperature, and humidity sensors.

Scott Pace designed and sewed together a parachute for Post 509's second big project, launching giant weather balloons. The balloons carried weather sensors more than 10 miles high and then burst. Scott's parachute floated the sensors back to Earth for recovery 10 to 15 miles from the launch site.

The programming team found a cheap way to store data in computer code: They used paper! A machine punched rows of holes in long, skinny strips in a pattern like this: no hole, hole, no hole, no hole, hole, no hole, no hole, no hole. A computer read that simple pattern as 01001000, which is computer code for the number 72.

~Hear Every Voice

Everyone was bursting with confidence and ideas. Then one day a key member of the electronics committee quit. Everyone got angry, including Adriana.

"How could he quit?" she asked. "He was the only one who knew how to build his component!"

The project was doomed without his circuit board. As the leader, it was Adriana's job to convince the teenager to come back. When she talked to him, she discovered that he was very frustrated and discouraged. At meetings he didn't think anyone was listening to his ideas, and he felt useless. He didn't see any point in showing up anymore.

Adriana understood. She realized that, as the leader, she had to keep everyone involved and motivated. She assured him, "Your voice will be heard at the meetings." After some user-friendly coaxing, he rejoined the team.

As the final review neared, another electronics whiz quit and Adriana was baffled. This teenager had worked hard and was very involved. Everyone liked her. Adriana couldn't imagine a reason for quitting, especially so close to the final review.

> She also clearly knew what she was doing. People trusted her judgment and respected her decisions.

Adriana and a few other Post members decided to pay her a visit. Their puzzlement and frustration quickly melted into sympathy. Adriana learned that the girl's mother had died, and the duties at home and at school had become hard to handle. The teenager had quit because she had no choice.

Adriana explained the situation to her team. She reminded herself not to judge someone's actions until she knew the person well and to take more time to get to know people. Scott Pace, an engineer to the core, thought of it as "unpacking people, just like you pry open a machine and unpack it to see how it works."

~ The Verdict

In April 1974 the EMC team was ready for the final review—but just barely. The whole team had pushed hard for months, getting ready for this moment. Adriana and Pat had worked all night to finish the report. The large von Kármán Auditorium was packed. Dr. Pickering himself was presiding, and Adriana felt both honored and nervous.

Adriana, age 19, delivers her EMC report to a packed audience in JPL's von Kármán Auditorium *(top photo)*. Dr. Bill Pickering, standing next to her in the bottom photo, was the director of JPL. In the late 1950s, he led the Explorer 1 project, the first successful U.S. satellite. A model of *Mariner 9* is behind him.

Her team took center stage, under bright spotlights in front of a plush curtain. As coordinator, Adriana stepped up to the podium and gave an overview of the EMC: The project was on budget, the ARMS ran well, and the APTRSS (now nicknamed "the albatross") was ready for construction. The smog detector was, to put it delicately, "in development." In other words, the team had met some goals but not all of them. If the parts that worked were deemed useless, the project was doomed.

JPL engineers and scientists toured the trailer and then rated the overall success. The verdict? "Funding approved." The EMC would live to see another sunny day!

Adriana was happy, relieved, and proud of her team, especially when a JPL engineer asked to use the EMC weather data for his radio project. She soon joined his team, called ARIES, and learned to build electronic circuits for the radio receivers.

~ Working on Mars

Adriana liked working with electronics, but in the mid-1970s, as she entered her 20s, she felt ready for a new adventure. In her rare free time, she drew designs for spaceships, took pilot lessons at

JPL's small airport, and still dreamed of flying into space someday. Meanwhile she decided to make some changes in her life.

She kept her $50 motorcycle but traded in her Karmann Ghia for a lively British sports car that was better suited to an adventure seeker like herself. Her green MG Midget was a tiny two-seater with a convertible top. It handled well, zipping through traffic with ease, and rode low to the ground.

Adriana also adopted a dog that *walked* low to the ground—a white and tan basset hound with typically short legs. The dog's name was Hermosa—"beautiful" in Spanish—but her beauty took a back seat to her role as Adriana's new copilot. Hermosa loved riding in the open "cockpit" of the car, her oversized ears flapping in the wind like flags in a hurricane.

What was to become the biggest change in Adriana's life involved a mysterious red planet, millions of miles away. One summer day in 1976, a tall, blond engineer named Tom Duxbury asked her, "How would you like to come work with me on Mars?"

To Adriana each image
from Mars looked like an
extraordinary puzzle,

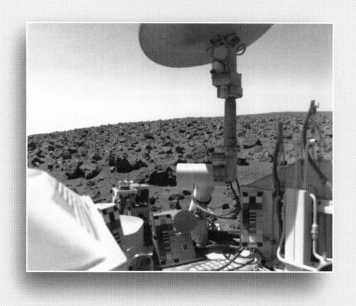

and she longed
to put the pieces together.

VISIONS OF MARS

5

When Tom Duxbury asked Adriana to work "on Mars," she knew he didn't mean on the planet itself. Robotic spacecraft, not humans, had that coveted job. Tom was asking Adriana Ocampo, age 21, to join the Viking project, JPL's biggest and boldest mission to date. In hallways and lunchrooms, it seemed like even the flies were buzzing about it.

On June 19, 1976, the whole JPL community cheered when *Viking 1* entered orbit around Mars. Its identical twin, *Viking 2*, would arrive in August. Both spacecraft were designed to split into two pieces, an orbiter half that would continue to circle Mars and a lander half that would park itself on the surface—in one piece, everyone hoped.

So far the success rate for a Mars landing was zero percent. The Soviet Union, America's powerful rival, had tried twice. In 1971 *Mars 2* and *Mars 3* arrived safely at the red planet, but the first lander crashed. The second one touched down, but lost contact with Earth 20 seconds later.

If either Viking lander succeeded, it would be a groundbreaking moment in space exploration. For the first time, Earthlings could see what it would be like to stand on Mars!

On her own time, Adriana volunteered to talk to students about women in space exploration. NASA lent her a car *(above)* and some spacecraft models. At JPL she was thrilled to join the Viking mission to Mars. *Viking 2 (opposite)* landed on a rock-littered plain called Utopia Planitia.

No one knew what was on that distant world, but Adriana had no trouble imagining it. She had read science fiction stories about watery canals, weird plants, ancient cities, and, of course, Martians! Yet images from *Mariner 9*, the first spacecraft to orbit another planet, showed no evidence of any of those fantastical things. Scientists were now wondering about tiny or hidden life signs: buried seas, fossils, and perhaps microscopic organisms beneath the frozen soil.

As Adriana entered Building 264, she hoped *some* sort of life form would crawl in front of the lander's cameras.

~ Robots Never Sleep

Inside the mission control center, she found that the Viking team was split up like the spacecraft: An orbiter team and a lander team worked in side-by-side rooms. There was a large conference room, lots of TV screens, and giant computers that took up a whole room of their own.

The lander half of the *Viking* spacecraft is a self-run science lab that's a little smaller than a car. JPL engineers tested this model in a California desert before parking twin landers on opposite sides of Mars. The long scoop *(lower left)* collects soil for instruments inside the machine to test for signs of life.

Right away Adriana sensed that she was part of something astounding. These busy people hoped to achieve things that others only dreamed about. Inside this bland office building, they were exploring space—and now, so was she! Though she hadn't left Earth, she felt as if her childhood dream had finally come true.

The Viking team numbered in the hundreds, but after the first spacecraft reached Mars, even that wasn't enough. Adriana and other young recruits were needed because robots never sleep.

Tom Duxbury, her new boss, explained, "We're taking images every day, all day." To make that fact sink in, he added, "That's 24 hours a day, seven days a week, 365 days a year. You can't let up."

Both Vikings beamed back a flood of computer code—long strings of 0's and 1's—that had to be translated into data and images. Adriana told Tom that she knew how to program the computers. Tom found that she was able to do that job and learn a piece of his job, too. Most days Tom's job was to figure out how to capture the two Martian moons—on camera, of course. The *Viking 1* orbiter had a digital camera along with several other data-gathering instruments.

Those moons aren't so easy to catch, Adriana noticed, as she watched him work. Deimos and Phobos were dark, very tiny, and speedy. Phobos, the inner moon, zipped around Mars three times in one Earth day. At the same time, *Viking 1* was speeding around in a different orbit. Meanwhile, Mars and its companions were zooming around the Sun. For brief moments, the Sun's rays, the Viking camera, and a Martian moon lined up favorably for a picture.

Plotting these "photo opportunities" and writing an orderly series of commands for the spacecraft to follow is called sequencing. The job is so complicated that each instrument on Viking required a dictionary-sized book filled with programming rules.

Tom Duxbury won awards for his work on Mariner, Viking, and the Soviet Union's PHOBOS mission to the Martian moon. With Adriana's help, he used Viking images to create the first Phobos atlas, a book of detailed maps.

~ *Flying Blind*

Adriana took a turn sequencing images for the next day's Phobos flyby. She couldn't see any of the flying objects—not even on radar—because they were tens of millions of miles away, traveling 20 times faster than jets. She felt like a blind air traffic controller. But air traffic controllers keep flying objects apart, at a safe distance from each other, whereas Adriana's goal was to bring objects close together in order to take good pictures.

This complex task took laser beam focus. Adriana forced all extra thoughts out of her mind. Then she used a computer to plot the next day's orbital paths for Mars, Phobos, and *Viking 1*. Noting the changing angle of the Sun, she chose the best point-and-shoot

An asteroid crashed into tiny Phobos, blasting out Stickney Crater on the moon's north pole and cracking almost the entire surface of the moon.

moments and wrote commands to send to the orbiter. Like any robot, the spacecraft would follow those commands to the digit. If there was a mistake, it was hers—not *Viking 1*'s. She double-checked her code and ran it through two computer programs to examine the newly plotted path.

To Adriana's great relief, "her" images came back the next day in focus and centered. She was thrilled. She had asked a space-traveling machine to do something—and it did it! That task had sounded simple, but she now knew how tough it was. In the first weeks of the mission, some photo ops had been missed. As the team got better at sequencing, the success rate rose to 100 percent. The two Viking orbiters would eventually send back nearly 52,000 images.

Tom's moon imaging team got an eyeful every day. One image showed an enormous crater on the north pole of little Phobos. The hole is almost half the width of the potato-shaped moon.

"An asteroid nearly blew Phobos to bits," Tom commented. He pointed out grooves over most of the little moon's surface. Some of the lines seemed to fan out from the crater. Adriana tried to imagine how they were made. *Did the force of the impact crack the rock? If so, did the cracks go down to the moon's core?*

~ Have a Safe Trip

Because lots of people were competing for the orbiter's camera time, Tom asked Adriana to "interface." He explained what that NASA term meant: "Our science team, the other science teams, the navigation team, and the spacecraft team—go and sit down face to face with them daily and work out issues." He saw immediately that Adriana was an excellent interfacer. She walked back and forth between the orbiter and lander rooms, listening, learning, and coordinating.

The lander team hoped to set down its half of *Viking 1* on July 4, 1976—the 200th birthday of the United States—on a site in Chryse Planitia. *Planitia* means "low plain," and in *Mariner 9*

> As a child Adrianita had favored science over safety while inventing her "killer" bug spray and launching paper airplanes from a rooftop. But now she understood what was at stake.

images the landing site appeared flat and smooth. But Adriana noticed that the team was looking at the new *Viking 1* images and frowning. She saw what was so troubling: Greater detail revealed that this "smooth" site was rough and uneven. If the lander set down on a boulder, the machine would tumble over and break. The engineers' main job was to keep *Viking 1* alive, so they deemed the site unsafe for landing.

Jim Martin, the Viking project leader, called a big meeting of scientists and engineers, and Adriana felt privileged to listen in. The two groups had different ideas about the best place to park the lander. The engineers wanted to choose a very smooth, flat spot in Chryse Planitia to increase the chances of a successful landing. The scientists argued for a site that was more scientifically interesting, even though a varied terrain had more hazards for landing.

As a child Adrianita had favored science over safety while inventing her "killer" bug spray and launching paper airplanes from a rooftop. But now she understood what was at stake. Although no lives were in danger, the landing was "all or nothing." Either the multimillion-dollar machine survived or the mission was over.

~ *Touchdown?*

Everyone finally agreed on a new landing site on the plain. On July 20, the anniversary of the *Apollo 11* Moon landing, the *Viking 1* lander separated from the orbiter and began plummeting through the thin Martian atmosphere.

Caltech's auditorium was packed with people—including Adriana's friends Pat and Scott—all watching the mission on a large monitor. JPL's von Kármán Auditorium was jammed with TV crews and journalists. Even Ray Bradbury, author of *The Martian Chronicles*, was there. But Adriana Ocampo, watching on a monitor in Building 264, had a front-row seat. She would be one of the first people on Earth to see the surface of Mars!

It was nearly 2 A.M., but Adriana wasn't about to go home. No one was. Finally, a few minutes before 5 A.M., a mission controller counted down the final feet as the lander descended: " . . . three, two, one, zero." The floor fell silent. Everyone had to wait 1,000 seconds to find out if the lander was still in one piece. Even at the speed of light, radio signals from Mars took about 18 minutes to arrive on Earth.

The first image of the surface of Mars arrived on Earth in a series of skinny vertical strips, from left to right. Viking engineers wanted a picture of the lander's footpad *(lower right)* right away to make sure the robot was standing firm and upright.

At 5:12 A.M., Jim Martin finally shouted, "Touchdown!" The whole JPL community erupted with cheers. The Viking lander was transmitting data from the surface of Mars!

Adriana watched a monitor, but it stayed blank while the radio signals traveled from lander to orbiter to Earth. Just before 6 A.M., the first batch of signals finally arrived, and a skinny vertical strip appeared on the left edge of the screen. Minutes passed as more signals streamed in and the screen filled, strip by strip. Shapes finally began to emerge in the fuzzy, black-and-white image.

Someone shouted, "Look! A foot!"

~ A Forest of Rocks

A foot? Adriana made out what looked like an upside-down pie plate. Then she got it. Earth's very first image from the surface of Mars showed the lander's own footpad!

Near the footpad, rocks came into focus. They had little pits. *What made those?* Adriana wondered. The JPL scientists began tossing around ideas: Water? Meteorite impacts? Dust storms with grains of sand that flew like little bullets? Trapped gas inside volcanic lava? The rocks also had a coating baked onto them. Adriana ached to flip one over to see if the coating was also on the other side.

The lander's camera revealed a stark landscape that someone described as a forest of rocks. Rocks of different sizes and shapes were strewn all over, their tops rounded by the wind. Between them were patches of sand dunes. *Why did those rocks scatter?* Adriana wondered. *Did the wind blow them across the surface? Or did water carry them long ago?* She couldn't take her eyes off the screen. She thought, *This seems way too familiar to be another world.* The Martian plain looked amazingly like the Mojave Desert in California.

The camera tilted higher to reveal a hill on the horizon. Then came a bigger surprise: The Martian sky was light, not inky black like the Moon's sky. The brightness was caused by sunlight bouncing off dust particles in the thin atmosphere.

To Adriana each image from Mars looked like an extraordinary puzzle, and she longed to put the pieces together.

"I really need to understand what is going on there," she told her friends. "I have to know what those rocks are about!"

The rocks were more than just puzzle pieces, though. As she listened to JPL scientists talk about the Viking images, Adriana began to see that the rocks were the only witnesses to a distant past. Written into their color, their shapes, their scars—every feature—was a story about what happened to them and to the planet they were on, long ago.

The image processing team, experts in turning computer data into pictures, added color to the Mars images. The Martian sky turned out to be light pink. Iron oxide (rust) on the surface makes the "red planet" appear orange.

The NASA scientists knew how to read some of the story, but there were hundreds of unanswered questions. Adriana wanted to answer them. At that moment she decided to change careers: Instead of aerospace engineering she would become a geologist!

~ From Utopia to Purgatory

On September 3, 1976, the *Viking* 2 lander touched down on the opposite side of Mars in Utopia Planitia. A "utopia" is a perfect or ideal world, but the Martian plain by that name turned out to be another lifeless desert. *Lifeless but fascinating,* Adriana thought. *These rocks are dusted with frost!*

What Is a Planetary Geologist?

The field of planetary geology—the study of rocks and land forms on other worlds—was born in the 1950s along with the Space Age. It exploded as a full-blown specialty of geology in the 1960s and 1970s, when robotic spacecraft began beaming back close-up images of the Moon and other planets.

The spacecraft showed that each world is unique. Mars is a frozen desert with a light pink sky. Venus is an inferno, hot enough to melt lead, with an atmosphere too thick to see through. Mercury is a dense, iron-filled ball with almost no atmosphere. Yet at the same time, every world follows the same laws of physics—gravity, heat, motion, and so on.

Planetary geologists are scientists who ask: What accounts for the differences? Why do these worlds look and act the way they do? They compare the geology of planets, moons, asteroids, comets, and other solid bodies.

Unlike the dry, dusty Moon, the surface of Mars has an occasional thin layer of frost (*left*) and polar ice caps that expand and shrink with the seasons.

In 1977, while working with the Viking team, Adriana finally earned her associate degree in science. She wanted to spend all her time "on Mars," but she needed a bachelor's degree, so she enrolled part-time at California State University, Los Angeles (CSULA). CSULA didn't have a program in planetary geology, which was a new field at the time, so she majored in geology and took a lot of astronomy classes.

Although a slowly rising number of women were joining her at JPL, Adriana discovered that she was the only woman majoring in geology during her first year at CSULA. By 1981 there were three. One professor told her that, in his experience, women weren't cut out to work in tough field conditions. He said that if she really wanted to be a geologist, she should set her sights on working in a lab, not in the field.

Adriana chose not to hear those words. She was hooked on solving geology puzzles—looking at a rock or land form and asking, "How did it get this way?" For that she needed to go outdoors, in the field, where there are rocks and land forms to look at. She was sure her thirst to learn was strong enough to lift her over any mountain or carry her across any desert.

She soon discovered that she was about to be put to that very test. To graduate she had to survive a trip to a scorching valley in Utah called Purgatory. In the Catholic religion purgatory is where souls go to endure a terrible punishment. Did the name fit? Adriana soon found herself faced with not one, but two painful ordeals.

If Adriana couldn't take
the heat and *pressure,*

*she went home without
a* college *degree.*

SCHOOL OF HARD ROCKS

6

In 1981 Adriana came close to losing her life. One night while driving home in her MG Midget, she saw out of the corner of her eye that a car was coming toward her. The car ran a red light and smacked her little convertible, knocking it into a light pole. Adriana had just enough time to duck. The other car landed right on top of hers, and she was trapped inside! She was crammed into a tiny space, unable to move and with her head throbbing in pain.

Emergency workers told her that they would have to cut her out using power tools. She tried to stay awake and alert as they worked for more than an hour. Incredibly, no bones were broken. The worst injury was a bad concussion, a severe blow to the head.

Adriana refused to let the accident darken her sunny outlook. She quietly recovered at her parents' home in South Pasadena—and then went back to classes at CSULA and work at JPL. Tom Duxbury, her boss on the Viking mission, saw that she wasn't done healing yet, and he was amazed at her toughness. She never complained or made excuses.

Adriana learned geology through textbooks, lab work, and outdoor field trips (*above*, hiking inside Barringer Crater). Her preference was clear: She had a passion for exploring rock formations, like the colorful layers of the Utah desert (*opposite*), in their natural settings.

~ The Pressure's On

Adriana had learned how to read soft rock layers like a history book, from youngest at the top to oldest at the bottom. The distinct features of each layer told her what Earth was like millions of years ago.

The Viking mission ended as Adriana was about to finish her geology degree. To graduate with the class of 1983, she had to pass two outdoor challenges: a hard rock test and a soft rock test. Hard rocks are igneous or metamorphic, two classes of rock that have been transformed by heat or pressure or both. Adriana passed that test by identifying and mapping a hard rock formation in California.

It was the soft rock test that landed her in Purgatory for six grueling weeks in the summer of 1982. Soft rock is the third class of rock, sedimentary, which is made from sediments (loose bits) that become cemented into solid rock. Purgatory Flat in southwest Utah had miles of colorful sedimentary layers that tourists found beautiful and Adriana found both scientifically and physically challenging.

Each day in Purgatory, she and her classmates camped in the desert, rose at dawn, hiked for miles, and studied rocks until mid-afternoon, when temperatures soared dangerously high. All that bare rock was heaven to a geologist, but it meant there were no plants to create shade. Adriana swallowed salt tablets so that her body would retain water and not dry out.

At a trim 5 feet 5 inches, she found that backpacking was her hardest challenge. She carried water, food, a hammer, and other heavy equipment. All of her muscles ached, and her legs grew more tired each day. No one dared to complain—especially the three women. Adriana wanted the professor to know she was as strong, or stronger, than the men. She recited the positive messages she had collected in her notebook and added one more: *Stamina!*

Adriana was searching for layers that didn't "pinch out"—thin out and then disappear. These long continuous layers of uniform thickness carry information about widespread events in the past—

like a major climate change. As she identified the layers by type of rock and relative age, she transformed the data into a geological map. If she could last the full six weeks and complete the map, she graduated. If she couldn't take the heat and pressure, she went home without a degree.

~ Geology Heaven

As her map began to take shape, Adriana noticed something interesting. Usually the oldest layers are at the bottom of a formation and the youngest ones are on top. Yet she was standing at the peak of a hill near an outcrop of very old rocks. *How did they get there?* Curiosity washed away some of her aches and pains.

Adriana's mind traveled back many millions of years, when this roasting desert

Adriana hiked around this colorful Utah desert in 1982. Above the lake, just right of center, is the rainbow-shaped anticline formation that she mapped *(inset)*. This arch of sedimentary layers dates back to the beginning of the reign of the dinosaurs.

Back in 1973, Adriana joined a Post 509 field trip. She remembers standing near the raised rim of Meteor Crater, imagining the prehistoric beasts that witnessed a space rock crash there 50,000 years ago.

was a refreshing sea. She pictured sediments streaming in from rivers and settling in a flat layer at the bottom of the sea. As new sediments piled on top, the lower layer became squeezed under the added weight and cemented into rock. Then as more sediments streamed in, another flat layer formed. And another layer. Over millions of years the sea turned into a blazing desert.

The layers of the hill she was standing on were still in order from oldest on the bottom to youngest on top, but they were no longer flat. The Earth's crust can rise, fall, shift sideways, and fold like an accordion. The layers beneath her feet, she realized, had been crunched together from either side, and the middle rose into a formation called an anticline. It was like squeezing a layered sandwich with both hands, forcing the center part to rise. As young layers at the top of the anticline wore away, the older rocks beneath them became exposed.

Adriana remembered the anticline diagram in her textbook, and now, suddenly, it made sense to her. *Standing on an anticline sure beats looking at one in a book*, she decided. The fieldwork had been dog-tiring, but Adriana thought the experience was closer to geology heaven than purgatory. She not only passed the test, she loved it! She couldn't wait for more!

~ Remembering Arizona

Adriana smiled as she remembered her first geology field trip 10 years earlier, a Thanksgiving vacation to Arizona with Post 509. What a contrast to Purgatory! Back in 1973 the biggest hardship had been losing her luggage and having to hike around the desert in dress shoes. Among other geology wonders, the Space Explorers had stopped at Barringer Crater, a hole so big that it made Adriana feel like an ant in a salad bowl—an ant with dress shoes on! Now, as a full-fledged geologist, she thought about that crater with renewed interest and curiosity.

In the early 20th century, some people called it a volcanic

caldera—a crater that forms after a volcano erupts and the ground collapses into an empty magma chamber beneath it. Then in 1957 a geologist named Gene Shoemaker found proof beyond question that Barringer Crater is not the slump of a caldera but rather an impact crater. A rock from space blasted out that hole!

In the 1950s the idea of a "killer" space rock was on the same shaky ground as flying saucers and little green men from Mars. Still, a few believers had dug for a buried space rock and never found one. Adriana remembered climbing into one of the many mine shafts in the wall of the pit. Now as a geologist, she understood why those shafts had all come up empty: Big asteroids hit with such explosive force that they melt or vaporize—turn to gas—in seconds.

Gene Shoemaker's conclusive proof wasn't in the asteroid; it was in the damage the impact caused to Earth's rocks. The force was so great that it had melted and shocked rocks in ways that no volcano was powerful enough to do. In seconds it had transformed quartz, a common mineral, into a very rare mineral called coesite.

Barringer Crater gained a second name, Meteor Crater, but the idea of space rocks hitting Earth took a while to catch on.

Adriana became friends with Gene and Carolyn Shoemaker, who loved to discover comets together. Gene proved that Barringer Crater in Arizona (above) is an impact crater and taught Apollo astronauts how to explore rocks and craters.

~ They Came from Outer Space

Space exploration put impact craters front and center in many people's minds—including Adriana's. On the Viking project, she had seen hundreds of impact craters on Mars and its two little moons. That whopper on Phobos still made her eyes pop. In 1979 images from JPL's Voyager spacecraft showed that one of Jupiter's moons, Callisto, had impact craters on top of impact craters, all over the surface!

At some point, any thinking person had to ask: Why *wouldn't* there be lots of impact craters on Earth, too? Clearly, they had to exist—but where? By the mid-1980s scientists were discovering a trickle of Earth craters per year—barely 100 in all. Adriana figured that many more were hidden somewhere, and she wanted to find them.

She knew why Meteor Crater was easy to spot. It was well preserved in a dry desert, and the asteroid had crashed only 50,000 years ago, during the last Ice Age. She thought about what kinds of beasts saw a bright streak cross the sky and felt Earth tremble. Woolly mammoths? Saber-toothed cats? Now after studying rocks hundreds of millions of years old, Meteor Crater seemed like a fresh-faced baby to her. To any geologist, thousands of years are like seconds and millions of years are like minutes.

> Rejection was disappointing but to Adriana, never a dead end. It just meant finding another way. In November 1988 Adriana Ocampo, age 33, was a woman with craters on her mind.

Adriana knew that most of Earth's craters, by far, were made at the dawn of the planet's existence. The craters had a few billion years to wear away or be erased altogether by wind, water, volcanoes, earthquakes, and other changes. Craters on the ocean floor could be swallowed up by the Earth itself! That happens when the crust subducts, or ducks beneath the neighboring crust into the molten rock below. Ocean craters that don't subduct become deeply buried under layers of sediment.

Adriana wondered, *How do you find something you can't see?* Her work at JPL sparked an idea: *Don't use human eyes to search. Use electronic eyes—the cameras on satellites.*

~ Craters on Her Mind

During the mid-1980s Adriana learned to process images from Landsat and Seasat, a pair of satellites in orbit around Earth. Instead of using visible light, Seasat took pictures using radar: It detected radio waves that bounced off Earth's surface and turned the wave patterns into black-and-white images.

Adriana hoped to use the technology to spot hidden craters. She figured that radar images from space could reveal the circular shape of a buried crater without digging or drilling. A satellite could also scan a wide area—the entire Earth, if needed. That was something people on the ground could never do. Adriana thought about how to tell a volcanic caldera from the blasted-out hole of an impact crater in a radar image. She didn't have an answer. It would take research to find an impact signature—a telltale clue that calderas didn't have.

She teamed up with other scientists to investigate the problems and possibilities. One was none other than Gene Shoemaker, the Meteor Crater investigator. Adriana had met him at JPL, where he sometimes lectured and advised NASA on space missions. The scientists wrote a plan and submitted it to NASA, hoping for money to fund the crater search. The proposal came back "not accepted." NASA management decided other uses for a space-based radar imager were more important.

Rejection was disappointing but to Adriana, never a dead end. It just meant finding another way. In November 1988 Adriana Ocampo, age 33, was a woman with craters on her mind. The last place she thought she'd find one was in a room full of people in Acapulco, Mexico!

A **killer rock** from outer space?
Wiping out entire life forms
with one big blow?

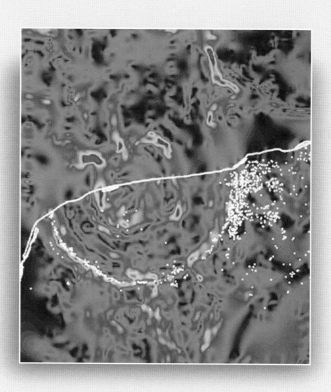

Most scientists
attacked that idea as **nonsense**.

Ring Around the Crater

7

Acapulco, Mexico, is a beach resort, but it was satellites, not sunshine, that lured Adriana there. She was attending the 1988 SELPER conference for scientists who use satellite images to forecast the weather, measure rain forests, map the oceans, spot hot volcanoes, and more.

As she wandered past the open meeting rooms, exploring her options, a Landsat image in an archaeology session caught her eye. She recognized the Yucatán Peninsula of southern Mexico. There was a blob of tiny black dots near the Gulf of Mexico shore. They were labeled *cenotes* [say-NO-tayz]—the Spanish word for sinkholes full of fresh water. Her eyes followed a trail of dots along a near-perfect curve down and to the left and then up again. It was a giant semicircle. She pictured a mirror copy extending into the water, creating a circle.

Land features don't form circles or semicircles by accident, she knew. So what created this ring of cenotes? After all her planetary adventures, Adriana's first thought was, *an impact crater!* She didn't want to be wrong, though, so her next thought was a more cautious, *impact crater?* If so it was deeply buried and bigger than almost every known crater on Earth.

A gravity map of rock density reveals a circular crater buried half a mile underground *(opposite)*. On the surface, a semicircle of sinkholes, marked by white dots, lines up above an outer ring. The white line marks the coast of the Yucatán Peninsula, Mexico. Adriana would later find ejecta *(above)*, material blasted out of the crater, a couple hundred miles away.

61

Identifying craters was still a new science, but Adriana wondered who else had thought about that cenote ring. After the session ended, she introduced herself to the speaker, Kevin Pope, a tall, sandy-haired Coloradoan who lived in northern California. As they chatted, Adriana discovered they shared a passion for fieldwork. In Kevin's case that meant wading through the wetlands of Central America and Mexico to dig up ancient Mayan ruins.

With his double degree in geology and archaeology, he dealt with young rocks—rocks that were just a couple million years old. He had no experience working with impact craters, but thought the cenote ring could be the result of a buried caldera. Mexico had plenty of volcanoes. The two scientists agreed to meet at JPL and look over the data that Kevin and his partner, Chuck Duller, had collected.

~ Their First Data

Adriana was surprised to see that someone else had spotted a giant circle in the Yucatán—decades ago. From the 1950s to the 1970s, Pemex, the Mexican petroleum company, had scanned the gulf looking for promising places to drill for oil. People can't see past the surface, but instruments can detect properties of the rock below. Two properties are magnetic fields (due to iron in the soil) and density (how heavy the rock is for its size).

A magnetism map showed a large circular anomaly [ah-NAH-mo-lee]—a patch of rock not like the others. Its magnetic field pointed in different directions than the rock around it. The gravity map, which plotted the density of rock, looked like a fragmented bull's-eye. An outer ring of dense rock ran directly under the arc of cenotes, which had

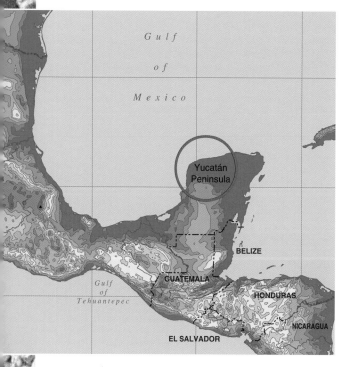

A red circle marks the location of one of the largest impact craters on Earth. When a space rock crashed there 65 million years ago, the world's sea level was higher. Mexico's Yucatán Peninsula and part of the southern United States, including Florida, were underwater.

formed much later. The whole structure was buried under about half a mile of younger rock that had piled up over millions of years.

To Adriana, the space explorer, these features were as familiar as old friends. "It's a beautiful example of what a large impact crater looks like," she said, her brown eyes shining.

She tried to picture the buried crater in her mind's eye. The exact size was hard to tell, but it looked *huge*. She remembered hobbling around the rim of Meteor Crater in her dress shoes back in 1973. The tour had taken a couple of hours, yet that hole was a pinprick compared to this monster. She measured the widest ring of the Mexican crater at 112 miles in diameter but thought it might be even bigger.

Meteor Crater's bowl shape and sharp rim are the marks of a minor impact. From images of Mars, Adriana knew that big space rocks make messy, complex craters, similar to this one in Mexico, because they crash with far more force. *How and when did it form?*

Adriana was hooked. Over the next few months, she and Kevin searched for more clues to this huge and mysterious crater.

~ Shocking Evidence

The two scientists learned that Pemex had drilled deep test holes in the underground structure, probing for oil. From the Pemex data, they saw that the samples contained several telltale clues of an impact.

One indicator was shocked quartz. When viewed under a microscope, this quartz has parallel lines caused by powerful shock waves. Three or more sets of parallel lines, criss-crossing each other, can only happen naturally when an asteroid or comet slams into Earth. Volcanic blasts, which are less forceful, create simpler patterns.

The thin, dark lines in this quartz are fractures (cracks) caused by shock waves of pressure. Fractures that criss-cross in three directions, as these do, are solid evidence of an asteroid or comet impact.

Adriana found this impact breccia among the Chicxulub ejecta. Ejecta includes any material tossed directly out of the crater during a collision.

Breccia [BRECH-ee-ah] is rock that contains large, sharp-edged chunks, like nuts and raisins in a cookie. Like shocked quartz, it can form in more than one way, but the Pemex sample was consistent with an impact. A third clue was the presence of tektites—rocks that melt and are thrown into the atmosphere. The droplets then cool into round or teardrop-shaped bits of glass and rain down over a wide area.

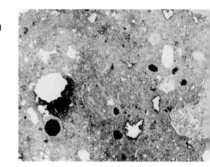

During the hunt for data, Adriana and Kevin discovered that other scientists had been hot on the crater's trail, too. These other teams had found layers of ejecta—material blasted out of a crater—in several North American sites. They had also found evidence of tsunamis [soo-NAH-meez], giant waves, on some of the Caribbean Islands.

Geologist Alan Hildebrand had followed this trail of disaster to ground zero—the point of impact. The space rock had crashed near a fishing village in Mexico named Puerto Chicxulub. (Chicxulub is a Mayan word pronounced somewhere between cheek-shoo-loob and chick-zoo-loob.)

~ Sudden Impact

After examining the Chicxulub crater, Adriana tried to imagine the massive explosion that created it. She pictured the impactor, a fiery ball traveling 100 times faster than a jet. Based on the crater size, she estimated its diameter at roughly six miles—wider than the Earth's highest mountains are tall!

Was it a comet or an asteroid? she wondered. A comet is a lightweight chunk of dirty ice and an asteroid is a dense, slower-moving space rock, but either one was possible. In a split second the impactor (the comet or asteroid) had exploded with a blinding flash and vaporized into a fireball of gas. This superhot vapor plume shot out in all directions, including up into space. Shock waves—powerful blasts of pressure—shaped the crater. Briefly, the hole was some 25 miles deep—as deep as the Earth's crust. But then earth rock that had melted on impact rushed down the sides, filling in the bottom.

More shock waves caused the center of the crater to bounce up and down. It rose a dozen miles high, up through the pool of melted rock. Big craters end up with a peak in the center, but Chicxulub was a giant. According to the gravity map, its center had rebounded again to form an inner ring with no peak—a sign of a *very* serious impact.

Even Adriana's imagination had to stretch to picture the global disaster that followed. The vapor plume, that hot fireball of gas, circled the entire globe. Plants and animals within about 600 miles were flattened by shock waves. The ground shook with earthquakes many times stronger than any on record. Wind faster than the worst hurricanes swept out from the point of impact, flattening trees like toothpicks. That left behind a vacuum, an airless space, which then drew the wind blast back in.

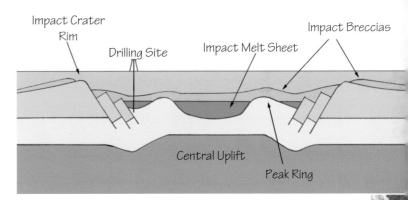

Adriana imagined the Gulf of Mexico as a bathtub, with tsunamis washing over the coastal rim again and again. Blobs of melted earth rock, blasted out of the crater into the air, rained down over a wide area and set wildfires. Forests burned, sending up thick clouds of soot and ash.

For many months there was utter gloom. Gases from the vapor plume blocked the Sun's heat, cooling the planet. Many weeks of total darkness prevented plants from making food and, as the plants died, so did plant-eating animals.

This was not a good time to be a resident of planet Earth. And Adriana had a pretty good idea who the unlucky beasts were. Soon after she had begun thinking "impact," a certain crater had stuck out in her mind because it was both huge and missing.

Huge impacts make messy craters. This simplified side view *(above)* shows the main parts of a complex crater like the one buried under the Yucatán, Mexico. The central uplift bounced up and down during impact, creating an inner peak ring *(beige)* that filled with melted rock *(red)*. Scientists had to drill through half a mile of rock to collect samples of the crater.

~ The Missing Crater

Adriana was friends with a famous geologist named Walter Alvarez. Back in 1980 Walter and his father, Luis, had flipped the science world upside down with a bold idea. They claimed they had

In Gubbio, Italy, Walter Alvarez *(right)* has a thin layer of dark clay at his fingertips. The clay divides the older, light-colored Cretaceous rock beneath his palm from the younger Tertiary rock that his father, Luis, is touching.

evidence that an asteroid or comet slammed into Earth 65 million years ago. Then their theory went further: This impact, they said, caused global changes that led to a sudden, mass extinction.

A killer rock from outer space? Wiping out entire life forms with one big blow? Most scientists attacked that idea as nonsense. They did agree there had been a mass extinction at the end of the Cretaceous period, 65 million years ago. More than half the species on Earth disappeared forever, including the dinosaurs.

The furor was over what killed these prehistoric creatures. For a century most scientists had pointed to a gradual cause, like a slow cooling of the planet's climate. Besides, they said, if the killer was a giant space rock, where was the crater it made? The hole had to be huge, and yet no one could find it. Proving an impact without a crater was like solving a murder without a body. It was possible, but it took a mountain of other, less direct evidence.

Throughout the 1980s, the investigation focused on a mysterious layer of sedimentary rock. In Gubbio, Italy, the Alvarez team found a band of dark clay, the thickness of a finger, neatly sandwiched between layers of limestone. The older limestone below the clay formed during the Cretaceous period, when dinosaurs ruled. It's rich with fossils, including tiny sea creatures called forams. The younger rock above the dark clay formed during the beginning of the Tertiary period. It has few forams, and many of the fossils are microscopic and malformed.

Clearly, something horrible had happened, and the secret lurked inside that dark clay. The Alvarez team discovered an unusually high level of a rare metal named iridium [ih-RID-ee-um]. Most of the Earth's iridium is inside the planet's iron and nickel core. So what was extra iridium doing in this thin clay layer on the surface? The scientists concluded that it came from outer space. A space rock containing iridium crashed into Earth, vaporized into a fireball, and then particles rained down all over the globe, forming that dark band of clay.

The clay layer lies at what is called the KT boundary, named after the older Cretaceous period below it and the younger Tertiary period above it. (The letter C already stood for the Carboniferous period, so K was used for Cretaceous.) Nearly 100 other KT boundary sites were found around the world, many of them by a Dutch geologist named Jan Smit. These sites contained shocked quartz, tektites, and impact breccias. One even had tiny diamonds that had been created instantly out of carbon by the great pressure of an impact.

By 1989 there was little doubt that an asteroid or comet had crashed on Earth 65 million years ago. Now Adriana believed there was an answer to that question, "So, where's the giant crater?" Chicxulub was the right age, the right size, and in the right location to be what Walter Alvarez called the Crater of Doom.

~ Hitched by a Crater

For Adriana and Kevin, Chicxulub became the Crater That Brought Them Together. As they reviewed all the data, the two began to date. On Friday, November 3, 1989, they were married at the Los Angeles County Courthouse. Victor and Teresa held a small reception at their home. The next day the newlyweds hopped on a flight to Africa.

It was a good thing they both loved fieldwork, since that was part of the honeymoon package. In Kenya, Kevin had a science project to do. He was using satellite data to help combat Rift

Who Discovered the Crater of Doom?

Giving credit for a big discovery isn't always easy. Taking credit isn't always fair.

The Chicxulub crater was first found by people who weren't looking for it. In the 1950s, Pemex, the Mexican oil company, found a large circular structure buried half a mile underground. But did anyone know it was an impact crater? Did anyone care? Pemex was far more interested in discovering oil.

Twenty years later two Pemex scientists, Antonio Camargo and Glen Penfield, rediscovered the old data and found new crater clues. While looking for oil, Penfield flew over the Gulf of Mexico with a magnetometer, an instrument for measuring magnetic fields. To his surprise, the instrument revealed a large circular anomaly—a spot not like the others. Pemex drilled test holes, still hoping for oil, and unearthed shocked quartz and other clues of an impact instead.

In 1981 the Pemex scientists announced they had found an impact crater. Journalist Carlos Byars wrote a story for the *Houston Chronicle,* but scientists who were looking for the very same crater didn't notice.

In November 1988 Adriana happened to spot the cenote ring in a satellite image at the SELPER conference. She and Kevin spent months trying to get their science report published. *Science* magazine turned them down, but *Nature* finally printed the story in 1991. During that time, geologists Alan Hildebrand and William Boynton had followed the ejecta and tsunami evidence to a Mexican village—Puerto Chicxulub—and had then reported their find.

So who gets the glory? Like a movie with endless sequels, the crater credits roll with hundreds of names: geologists, physicists, chemists, paleontologists (fossil experts), botanists (plant experts), biologists, and other explorers who helped put together the pieces of the puzzle.

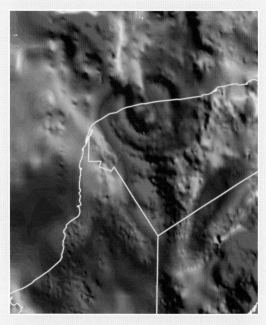

This gravity map plots the density of buried rock from magenta and blue (low density) to red (high density). (Denser rocks have a greater pull of gravity.) The magenta circle is a low-density ring of Chicxulub crater. The blue patch in the center is slightly higher-density rock that melted on impact and flowed into the hole.

Valley Fever, a deadly illness. The images revealed wetlands where disease-carrying mosquitoes were likely to live.

The couple did have time for a safari to the Serengeti Plain in the Masai Mara National Reserve. Moses, the driver, told them that as long as they stayed inside the vehicle, they could drive right up to the wild animals. "If you step outside, they will either run or eat you."

Sure enough, zebras, wildebeests, rhinoceroses, and antelopes grazed near their little bus, paying no attention to them. The animals' minds were on a bigger threat.

"Everything is peaceful until the lions get hungry," Adriana observed.

After Kenya the couple flew to Argentina for the annual SELPER meeting and to introduce Kevin to Adriana's relatives and friends. Some couples have "their" song, a tune that brings up warm memories. Kevin and Adriana had a crater. In the spring of 1991 Chicxulub would provide them with the scientific discovery of their lives.

Investigating an incredible site
like this would take *years of digging,*

lab work, mapping,

and **debating** the data.

HEAVY IMPACT

8

For both Adriana and Kevin, finding one piece of a puzzle made them want to chase harder after the next one. A trail of clues had driven them from a mysterious ring on a satellite image to an impact crater to the Crater of Doom to . . . Wow! What's *that?*

Whatever "that" was had to wait for now. After identifying one of the largest impact craters on Earth, it was hard to put on the brakes, but Chicxulub was deeply buried. Drilling through a half mile of rock was a multimillion-dollar task best left to oil companies and big universities.

In the spring of 1991 Kevin's archaeology project took him to Belize, a tiny country tucked south of the Yucatán Peninsula. In the northeast, near the Mexican border, he was digging in a swamp for Mayan ruins. Adriana was in California, working on JPL's Mars Observer team, but the launch was postponed. She decided to join her husband. Tropical heat and mosquitoes were a small price to pay for being able to play in the mud.

The two scientists weren't lie-on-the-beach travelers, so on a free Sunday they found a different way to relax. They decided to look for a KT (Cretaceous-Tertiary) boundary site. In a land of swamps, sugarcane, and jungle, finding exposed rock wasn't easy, but Kevin knew about a rock pit near a sinkhole where he often swam.

Chicxulub crater was a gigantic hole created in seconds by a crashing space rock. What happened to the Earth rock that was inside that hole? Adriana was shocked to discover that massive amounts of ejecta *(above and opposite)* traveled more than 200 miles away in a matter of minutes!

~A Closer Look

The border between Belize and Mexico is a river—the Rio Hondo. The rock pit is on Albion Island, a stretch of land surrounded by the river and the Pull Trouser Swamp. Kevin and Adriana drove across a bridge and parked in the bottom of the pit. They and their vehicle were dwarfed by cliffs as tall as a 13-story building.

So far all the KT boundary outcrops had been measured in fractions of an inch, so Adriana and Kevin walked closer to the tall wall of rock. The base was neatly layered limestone, one even band on top of the next. About three-fourths of the way up the wall, they spotted a distinct orange layer about three feet thick.

Adriana climbed up the cliff and pulled out her hammer—no geologist leaves home without one—and chipped off samples of the orangish rock. She smiled. It was breccia. The "nuts and raisins" in this limestone cookie were smooth and rounded and glassy. They were cherry sized or smaller and came in orange and

On Albion Island, Belize, workers harvest limestone, a sedimentary rock, for use in construction. Note the tiny-looking trucks at the base of the towering cliff.

pink and white. She called them spheroids—little ball-shaped objects. Mixed into the orange layer, at the bottom, were other balls of shiny green clay.

It's easy to jump up and down and shout, "Eureka! I found it!" It's very tough to prove scientifically that you're not just flapping your jaws. Adriana suspected that these spheroids had come all the way from Chicxulub, but there were questions: Could a volcano have formed them instead? If they were from an impact, was the rock the right age for Chicxulub— 65 million years old? She noticed that not all of the fragments were round and smooth; some were squashed or chunky or coarse or lumpy. How did they get that way?

At the Albion Island rock pit, Adriana spotted that distinctive orange layer right away. The rock above it contains a jumble of mudballs *(below)*.

~A Big Step Back

Adriana hammered at the rock on top of the orange layer. It was a jumbled mix of rock pieces, sand, and mud that had turned into solid stone. Many rock pieces looked fractured and faceted, with smooth faces cut into them, like those on a gemstone. Some had polished areas and rounded edges and grooves that ran in different directions. Adriana broke open a chunk and found that it was a mudball— a rock coated with a layer of limestone mud, like a nut rolled in chocolate.

All of these features were like a written record of what had happened to the rocks in

Adriana wondered if the crazy mix of mudballs—rocks coated with limestone mud—and sharp-edged boulders were ejecta from the Chicxulub crater. She photographed her trusty hammer against one of the mudballs as a scale reference.

Adriana examines the Albion Island rock pit for clues to what happened there 65 million years ago. Did the rocks, some of them the size of cars, blast out of the Gulf of Mexico and land on Albion Island within minutes?

the past. As Adriana's friend Walter Alvarez liked to say, "Liquids and gases forget, but solids remember." From the marks and scars, Adriana understood that these rocks had been through some serious and brutal trauma.

She dared to think, *Like the orange layer, could this layer also be ejecta from the crater?* Chicxulub was 225 miles away! She stared at a boulder the size of a car. *Is it possible?* She stepped back and took in the big picture. The jumbled top layer and the orange band of spheroids beneath it were continuous. They ran together for yards and yards and yards around the pit.

Adriana's heart pounded. *Could this be the thickest KT boundary layer ever found? Right here, out in the open?* She wanted to shout, but she told herself to take a mental step back. Again she knew better than to state any science "fact" without plenty of data to back it up, and so did Kevin.

As they headed back to the swamp, they knew what lay in front of them. Investigating an incredible site like this would take years of digging, lab work, mapping, and debating the data.

~How It Got That Way

From 1991 to 1994 Adriana and Kevin visited the Albion Island rock pit whenever they could, working around his archaeological digs and her space missions. They invited along other scientists, including Al Fischer, a geology professor who had taught both Walter Alvarez and Adriana. Al confirmed their conclusion: This was indeed Chicxulub ejecta of the most extraordinary kind. It was part of the ejecta blanket—the deposits flung directly from a crater that form an unbroken sheet all around it.

Adriana imagined the event in her mind. She pictured the impact, the explosion, the gas fireball, and these boulders flying out of the crater like oversized bullets. Meanwhile, a curtain of melted rock was rolling over the ground like a liquid avalanche. The rock pieces fell into this curtain and were tumbled and scraped and polished and faceted along the way.

Al estimated that the rocks had arrived at Albion Island only 15 to 18 minutes after impact! That's like zipping across Missouri, from Kansas City to St. Louis, during a snack break.

To Adriana, the more they found, the more there was to find. How far into Belize did this ejecta blanket go? Was it similar to ejecta blankets around craters on Mars, some of which had lobes, like the petals of a flower? What about the orange spheroids underneath the jumbled rock pieces—how did they arrive before the ejecta curtain rolled in? And why were they squashed?

Adriana and Kevin needed a small army to help search the site for answers to that basic geology question, "How did it get that way?" This job is called field*work*, but Adriana figured out that certain adventure seekers might volunteer—and even pay money for the privilege of helping.

Adriana *(right)* and her volunteer team study satellite images and maps in the search for more crater ejecta. By the last expedition, in 2001, Planetary Society crews had helped locate and explore half a dozen KT boundary sites in Belize and southern Mexico.

~The Planetary Crew

The tropical country of Belize is well covered in swamps, forests, and sugar-cane fields, so an outcrop of exposed rock like Pook's Hill *(below)* draws a geology crowd. Planetary Society volunteers are learning what rocks to look for.

The Planetary Society is a group that promotes space exploration. The members had a keen interest in NEOs—near-Earth objects—the space rocks that could devastate our planet in the future. Adriana thought they might want to learn about an NEO that had already slammed into Earth. The Planetary Society's director, Lou Friedman, agreed to see how many members would pay to chip away at a rock pile. *A lot*, it turned out.

The society sponsored four expeditions to Belize in 1995, 1996, 1998, and 2001 and one to Gubbio, Italy, the site of the KT clay layer made famous by the Alvarez team. Lou Friedman understood the importance of this fieldwork. He explained to a member, "When you have a controversy, *data, data, data* answers everything. Go out and get data! That's what the real scientist does—and Adriana is always willing to go out and get data."

Adriana and Kevin led all five trips, and Al Fischer, Walter Alvarez, Jan Smit, and many other top scientists took part. The Planetary Society volunteers couldn't wait to "dig in." Sandy Miarecki [my-RECK-ee], a fighter jet pilot, said, "We were the worker bees,

Volunteer Sandy Miarecki *(right)* helps Adriana gather samples of ejecta rock at 10-centimeter intervals. This KT boundary layer, in Quintana Roo, Mexico, is as tall as a four-story building.

On the Trail of Global Disaster

Geologist Jan Smit has studied more KT (Cretaceous-Tertiary) boundary sites than anyone in the world. Like Walter Alvarez, he began piecing together the mass extinction puzzle in the late 1970s. The impact was a global disaster, so it's helpful to step back and see the big picture.

With a worldwide view, Jan says that as you near the Gulf of Mexico, "you definitely get the impression you are zooming in toward ground zero, the center of the catastrophe." He started out studying KT layers in Spain and Denmark that are only a fraction of an inch thick.

"The impression is that everything is quiet and not much energy is involved," he explains. But then in the western United States, the ejecta layer becomes 10 times thicker—and thicker still as you move south. Near the Brazos River in Texas, Jan found an amazing KT boundary with half a meter of sandstone.

"[It's] the first sign that indicates lots of energy was spent in the environment," he says. "Huge waves up to 100 meters high crashed on shore."

Farther south, in Mexico, the layer thickens and the evidence becomes more dramatic: "Whole parts of the continental margin collapsed, due to ever-increasing energy," Jan explains. Earthquakes created giant piles of rubble, riddled with small tektites thrown high into the air by the impact.

"At last, in Belize, the first signs of the so-called ejecta blanket become visible," he continues. Unlike in Mexico, there are huge boulders that flew or flooded into the area.

"From images of Mars and the Moon, we know that such ejecta blankets never extend farther away than two or three times the radius of the crater," Jan says. "Thus we finally know we are approaching ground zero."

Like Chicxulub, Rampart crater on Mars formed in the presence of water. The telltale clue is the fluidized ejecta blanket—ejected rock that flowed over the surface like a liquid. Lobes of ejecta fanned out from the crater in all directions, like the overlapping petals of a flower. Adriana's Belize research will explain how that process happened on Mars.

looking, digging, and carrying samples." Afterward, there were science talks, tourist sights to see, and even a talent show.

The "worker bees" helped collect nearly half a ton of rock samples in the Albion Island pit and at several other KT outcrops in Belize and southern Mexico. The scientists debated the data. Jan Smit and Adriana talked about that orange spheroid layer. Adriana thought the spheroids shot out along with the gas fireball, which overran the slower-moving liquid curtain that contained the rock pieces. The gas would have passed over Albion Island a few minutes after impact, raining down little spheroids. Then the curtain of melted rock flowed in on top of them.

Jan wasn't so sure. He thought that if the melted rock had run over the orange spheroids, they wouldn't be so evenly spread. The spheroid layer would be more disturbed and patchy. Instead he pictured the spheroids traveling *under* and *with* a liquid curtain of rock, not above it and ahead of it. The spheroids would have formed a lubricating (slick) layer between the melted rock and the solid ground, and the balls became rounded through milling, or constant grinding.

~ Pook's Pebbles

Adriana crosses the Rio Hondo, the river between Belize and Mexico, on a car ferry. She found crater ejecta on both sides, often by accident. One rainy day her vehicle slid off the road and got stuck in the mud—right next to an outcrop of Pook's pebbles.

One rainy, gloomy night, while the scientists and volunteers searched for more KT boundaries, the Rio Hondo ferry broke down. They were forced to drive back toward the capital of Belize, Belmopan, and find a place to stay. Traveling in darkness along poor roads, Adriana drove into Teakettle Village and saw a tiny sign that said "Pook's Hill Lodge." She went for it. The turnoff zigged and zagged up toward a remote hideaway near the Maya Mountains.

The next morning Adriana woke up to a science surprise. The lodge turned out to be set in a jungle paradise, but what drew her eye was an outcrop of rocks. The outcrop contained pebbles that

were roundish, pink, and smooth, like small stones on a beach. Some had pits—tiny, rimmed craters—from high-speed impacts with smaller objects. The group nicknamed them Pook's pebbles, after the lodge.

How did they form? Were these crater ejecta, too? Adriana thought so, but it was impossible to tell from just one site. Luckily, a second outcrop stared her in the face not that long afterward—again by accident and again on a chilly and rainy day. This time Adriana was riding in the car with Kevin. They slowed down at a speed bump right before entering the town of Armenia, and Adriana happened to glance out the window. There it was: a cutaway dug into a hill to let the highway pass through. It had a long stretch of large, colorful layers that looked like crater ejecta. No one wanted to get out in the rain, especially not Adriana, who was feeling sick. But this site was too good to pass up.

At Pook's Hill Lodge *(left)*, Adriana found an outcrop of strange pebbles near her hut. Some Pook's pebbles, as they were named, have tiny pits *(below)*. Were the pebbles struck by high-speed particles as they flew inside the fireball from the Chicxulub impact? The black-and-white marker is one centimeter long.

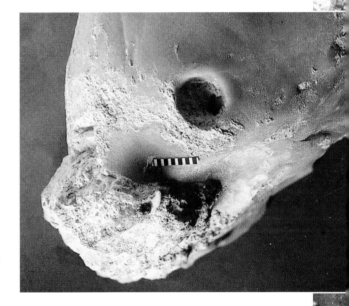

~An Ejecta Mystery

Standing in the rain, Adriana read the layers from oldest to youngest. The oldest layer was the same as the neat limestone base at the Albion Island rock pit. It formed in the late Cretaceous period, when dinosaurs still existed. The next layer began with a clear KT boundary—an abrupt change to spheroids. The youngest layer, in direct contact with the spheroids, included Pook's pebbles!

Pook's Pebble Bed

Spheroid Bed

Late Cretaceous Rock

When workers in Armenia, Belize, cut into this hill to let the highway pass through, they exposed stunning layers of crater ejecta. The organized layer on the far right formed before the Chicxulub impact, in the late Cretaceous period. Then, 65 million years ago, a thick jumble of ejecta arrived and piled up here in minutes. Did the Pook's pebbles (on the left) arrive as part of the vapor plume? Or did they form later? It's a mystery.

Adriana and her team had just discovered the farthest direct ejecta from the Chicxulub crater—and that fact presented a mystery. How did so much material travel that far? One idea was that the ejecta didn't come from the main crater but rather from secondary craters. In other words, the Chicxulub impact blasted out huge chunks of rock that went flying, crashed, and formed their own craters with ejecta blankets. Another possibility was that the material traveled as part of the gas fireball, which moved faster and farther than the other ejecta. Adriana pictured Pook's pebbles raining down from the hot gas cloud like hailstones.

Kevin and Jan weren't completely convinced yet that the pebbles were ejecta, since round balls of limestone can form in other ways. For example, a mineral that's dissolved in water can precipitate—drop out of the liquid and form a solid. The solids become smoothed and rounded by being washed over by waves and by being ground against other rocks.

~A Changing Picture of Doom

Kevin and Adriana worked with a JPL colleague named Kevin Baines to create a model—a computer calculation of what happened after the impact. Among the data they entered into the computer was a number that would turn out to change our picture of the mass extinction. That number was the huge volume of sulfur, a harsh chemical, buried with the limestone under the Gulf of Mexico.

The computer showed that heat and shock pressure would have vaporized the sulfur and limestone and injected massive amounts of both gases into the upper atmosphere. With so much "killer" sulfur in the air, the scientists realized that the dinosaurs had bigger problems than the clouds of dust and soot that blackened the sky. The sulfur combined with water vapor to create terrible acid rains, the kind that eat away vegetation. On top of the global fireball and months of blocked sunlight, the acid rains made a mass extinction even more massive.

In the meantime, while investigating the deadly disaster recorded in KT rocks on Earth, Adriana had continued to explore lifeless moons and planets at JPL. But were these other worlds *really* lifeless? For a little icy moon named Europa, the answer was changing from "yes" to "maybe not."

Adriana **sequenced**
more than a dozen flybys of Europa,

*and her images led to some
of the mission's most amazing*
discoveries!

9

SPACE, REVISITED

Adriana Ocampo was living a busy life. While she searched the Earth for crater ejecta, she was also exploring the solar system through the electronic eyes of spacecraft. She was, after all, a *planetary* geologist.

In the early 1980s she had worked on the navigation team for *Voyagers* 1 and 2. She helped the twin spacecraft steer toward Saturn and loved the idea that, like ancient sailors, they relied on the position of the stars to find their way.

In 1984 she joined the Galileo mission. That Jupiter-bound spacecraft flew in and out of her life for the next 14 years. Galileo, the great Italian scientist, had discovered Jupiter's four largest moons in 1610. Through his early telescope, they looked like tiny, identical points of light, but the Voyager mission showed that the moons are in fact as different as fire and ice. *Galileo*, the spacecraft, would orbit Jupiter for a more detailed look.

Adriana was one of about 40 scientists and engineers assigned to a *Galileo* instrument called NIMS: Near-Infrared Mapping Spectrometer. With near-infrared vision, a NIMS camera can peer through thick clouds, just as an X-ray machine can see through flesh. It is also a spectrometer, an instrument that identifies chemicals in an atmosphere or on a surface.

In 1989, Adriana took another tour of the launch site at Kennedy Space Center *(opposite)*. This time, instead of Apollo astronauts, a robotic spacecraft named *Galileo* was launched into space aboard a shuttle. Adriana and her teammates took a break in Hawaii *(above)* during what stretched into a 14-year mission.

In 1985 *Galileo* was shipped off to the Kennedy Space Center to await launch aboard a space shuttle in May 1986. Unfortunately the Jupiter-bound spacecraft never made it past Florida.

~ The Long Road Home

On January 28, 1986, the space shuttle launch prior to *Galileo's* ended in disaster. Seventy-three seconds after liftoff, *Challenger* exploded.

Adriana heard the shocking news as she arrived at JPL. She went directly to the von Kármán Auditorium, where a stunned and silent crowd was glued to TV monitors. To Adriana it seemed even the birds had stopped singing.

"Obviously a major malfunction," said a voice on the TV.

A video of the explosion played again and again. Then the reality of it sank in: The space shuttle was gone. All seven astronauts were dead. Adriana didn't know them, but the five men and two women were space explorers, like her. She felt a kinship with them.

All future launches were suspended. NASA decided that *Galileo's* Centaur rocket, powerful enough to propel the craft to Jupiter, was too dangerous to put on a shuttle and canceled the mission altogether. In Florida instruments were plucked off the $700 million machine and sent back to their makers across the United States and in Germany. The stripped-down skeleton was shipped back to California in 1987, a sad sight for the JPL community.

A video of the explosion played again and again. Then the reality of it sank in: The space shuttle was gone.

Galileo was a metal mess, but the team didn't give up on the mission. To launch the spacecraft safely on a shuttle, the engineers abandoned the Centaur rocket and came up with another way to send *Galileo* to Jupiter. A rebuilt spacecraft would get a free ride using the gravitational pull of Venus and then that of Earth. Each planet would pull *Galileo* toward it, speeding up the spacecraft, and then *Galileo* would whip around the back of the planet and be flung on its way at a higher speed.

Visiting Venus and Earth before heading for Jupiter meant a winding path that would take six years, instead of two. The new launch date was set for the fall of 1989.

During 1989, Adriana Ocampo had her eyes on a lot of prizes. At age 34, she was studying for her master's degree in geology at California State University, Northridge. (She would graduate in 1997.) She was also working on Mars Observer, the first U.S. mission to the red planet since Viking. (Sadly, this spacecraft would crash on arrival, prompting Adriana to comment, "It's very challenging to orbit another planet.") Meanwhile, she and Kevin continued to chip away at rock in Belize whenever they could.

And then there was *Galileo*. On October 18, 1989, the space shuttle *Atlantis* finally lifted the spacecraft off the surface of Earth!

On October 18, 1989, Adriana snapped this picture of the space shuttle *Atlantis* lifting *Galileo* off the surface of Earth.

~Is There Life on Earth?

Over the next decade *Galileo* had a series of space encounters and, during each exciting event, Adriana worked overtime in Building 264 at JPL. As a science coordinator, she did a lot of sequencing, plotting a series of images for the NIMS camera to shoot.

In December 1990 she sequenced *Galileo*'s flyby of Earth. Our home planet was already well scanned by satellites, but Carl Sagan, the famous astronomer, proposed a "Search for Life on Earth."

Adriana thought it was an interesting idea: From space, Earth is a mostly blue ball with swirls of white clouds. Forests and cities are too small to see except from a very low Earth orbit. She wondered, *How could you be sure that life was down there, if you didn't already know? What signs would you look for from space? Where on Earth would she point the NIMS camera to search for life?*

Working with Carl Sagan and the Galileo science team, she plotted out her sequence of images. NIMS spotted lots of water—oceans, ice, and water vapor—almost everywhere on Earth that it looked. Water is necessary for life but doesn't prove its existence. Instead, the instrument's best clue turned out to be a gas. The gas wasn't oxygen (which plants produce) or carbon dioxide (which animals exhale). Those gases could have nonlife explanations. The telltale gas was *methane!*

Methane quickly breaks down into carbon dioxide and water, and yet the NIMS spectrometer showed that the atmosphere had tons of methane. Something was producing the gas at a faster rate than it was breaking down.

"Life!" was Carl Sagan's conclusion. Bacteria produce about half of Earth's methane, and humans make most of the rest through agriculture and the burning of fossil fuels.

Adriana checks the "health" of the NIMS instrument aboard *Galileo* by monitoring data on a computer screen *(top photo)*. Weighing just under 40 pounds, NIMS *(above)* used 12 watts of power on average to do its work.

In October 1991 Adriana had a front-row seat at JPL as *Galileo* became the first spacecraft to fly by an asteroid. That was just a warm-up, though. In 1993 *Galileo* flew by another tiny space rock named Ida that has its own, even tinier, moon!

In July 1994 a rare and stunning space event prompted Adriana to take her whole family to an observatory to see it. Astronomers around the world were tracking a string of 21 fragments that used to be comet Shoemaker-Levy 9. (Adriana's friend Gene Shoemaker and his wife Carolyn loved to discover comets.)

The comet pieces were on a collision course with Jupiter's far side, and the space-traveling *Galileo* had a clear view of the impact. Over five days the spacecraft snapped pictures as the comet chunks disappeared into Jupiter's atmosphere, one after the other, in puffs of brown vapor.

Adriana wondered about a similar series of impacts on Earth. Maybe the Crater of Doom had siblings!

~ Jupiter, At Last

On December 7, 1995, *Galileo* finally entered orbit around Jupiter. Building 264 exploded with activity, and Adriana was in the thick of it. The spacecraft kept whizzing past the moons Io, Ganymede, Callisto, and Europa—in changing order. The four NIMS science coordinators, including Adriana, had each picked a moon to study.

One of the coordinators was Rosaly Lopes, a volcano geologist from Brazil, the country just north of Argentina. Since joining the Galileo team in 1991, she had become close friends with Adriana. Rosaly chose Io, a volcano lover's dream.

During its 1979 flyby, *Voyager 1's* camera had captured a volcano on Io in the act of erupting, and Rosaly thought she'd be lucky to find a few

An artist imagined *Galileo* arriving at Jupiter. The blue dots stand for data being received by a probe dropped into the planet's atmosphere. The real *Galileo* lost the use of its main antenna when its orange umbrella *(above, left)* didn't open properly.

Galileo's images of Io captured a tiny bluish smudge on the moon's far edge—a plume of gas from an erupting volcano. The plume is 86 miles high!

In *Voyager 1* images taken on the fly, Europa looks like a smooth, white billiard ball *(right)*. Adriana programmed *Galileo* to take a much closer, detailed look at its icy surface. In July 1995 she invited her father, Victor, to JPL *(below)* when *Galileo* sent a probe into Jupiter's atmosphere. The billboard charts *Galileo's* long, loopy path through space.

more hot spots. Over the span of the *Galileo* mission, she would use NIMS to discover 71 active volcanoes, some of which are far hotter than any on Earth! The colorful little moon was constantly redecorating itself with fresh lava.

"I think I got the best moon," Rosaly said, smiling.

Adriana felt the same way. She had chosen Europa because, with its crust made of ice, it was unique in the solar system. In the *Voyager 1* images, Europa looked like a fuzzy, pale billiard ball. The smooth surface was slashed with dark lines, like a modern abstract painting.

Adriana knew *Galileo* would shed plenty of new light on this mysterious moon. From 1996 to 1998 she sequenced more than a dozen flybys of Europa and her images led to some of the mission's most amazing discoveries. A team of scientists would be analyzing the images that Adriana sequenced for years. As a planetary geologist herself, she couldn't help but peer down at the icy surface of Europa and ask: *How did it get that way?*

In 1993, *Galileo* flew through the asteroid belt between Mars and Jupiter. Close-ups of Ida, a space rock *(right),* revealed that little round surprise on the right— the first moon ever discovered around an asteroid. Tiny Dactyl, about a mile wide, was likely blasted off Ida's surface during a collision with another asteroid.

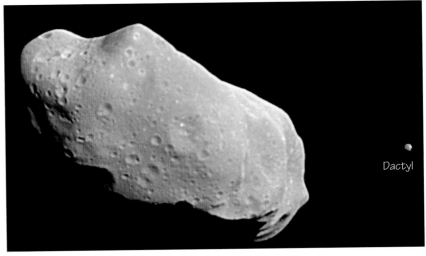

Dactyl

~ A World of Ice—and Water?

Europa's icy crust is criss-crossed with long, straight ridges, some with triple bands. In one place ridges frame large, squarish sections of ice, like lines on a colorless checkerboard. Adriana wondered, *Why the long, straight ridges? Why geometric squares of ice?*

She thought the pattern looked similar to Arctic ice, which floats on an ocean. As the ice sheet melts in the spring, long cracks form and then rafts of ice jostle and bump together. The rafts refreeze into a solid sheet in the fall, but not a smooth one—the battered edges become ridges. *Did Europa's crust once have "ice rafts" floating on an ocean?*

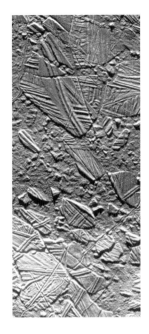

In the Conamara Chaos region of Europa *(left)*, the white spot is a patch of ice particles ejected during an impact (the crater is to the south, out of the picture). The brown material is a dusting of minerals. Rectangular ice rafts *(lower photo)* may have broken apart at the seams and then refroze.

One of the most stunning Europa images showed an area called the Conamara Chaos. Adriana studied the crazy jumble of criss-crossed lines, scribbles, scars, and ridges. The surface was frozen

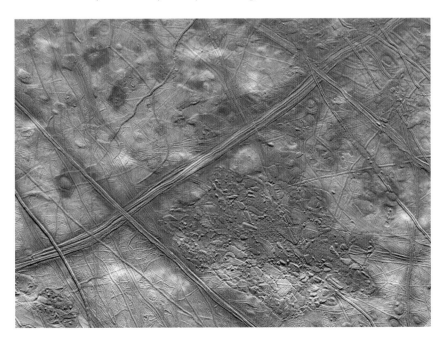

solid, but she saw signs that it, too, had probably cracked apart and floated in the past. Some chunks looked as if they had spun halfway around—a rotation possible while on water—before refreezing again. There were soaring ice cliffs, perhaps the result of floating blocks crashing together and flipping onto their sides. In some places it appeared as if water had splashed onto the ice from below and frozen into smooth ponds of ice.

This evidence for a past ocean was strong, but was Europa *still* a water world? Adriana knew that direct proof would require a lander to drill through the icy crust. Other evidence could come from a temporal (time) study in which a spacecraft flies over the same area again and

> None of this was *proof* of life, but if life *had* formed on Europa, it was likely still there.

again. Scientists then compare each image and look for a change due to movement, such as ice that floated. Io's volcanoes were so active that changes happened from one flyby to the next. Frozen Europa would require a much longer temporal study than *Galileo* could provide.

~ Carbon-Based Life Forms?

Adriana had read Arthur C. Clarke's *Space Odyssey* series of novels in which he predicted Europa had an ocean with life forms swimming in it. The first part seemed to be turning from fiction to fact. *What about the second part?* Adriana wondered. *Did a European ocean once harbor life? Does one harbor life now?*

She knew that living organisms can thrive where there is liquid water, heat, and the element carbon. NIMS had revealed that the mysterious dark streaks are made of minerals that formed while soaked in liquid water. The presence of liquid water required heat to keep it from freezing.

What about carbon, the third ingredient for life? Data from *Galileo's* instruments suggested that the streaks of minerals consist of white salts colored with other minerals that are red and yellow—perhaps a dusting of sulfur from Io's volcanoes. But they might also contain carbon-hydrogen and carbon-nitrogen compounds.

None of this was *proof* of life, but if life *had* formed on Europa, it was likely still there. Adriana explained to a reporter from EarthSky Radio: "The amazing thing is, if you get life started, it takes a tremendous amount of energy to stop it. It takes an incredible catastrophe to completely wipe out life."

The Galileo scientists believed strongly enough in the possibility of life to destroy their spacecraft! They feared that if the machine ran out of fuel, it might crash into the moon and pollute its surface with Earth microbes, making future searches for life difficult. Partly to keep Europa untouched, the team sent *Galileo* on a self-destruct mission into Jupiter's atmosphere in 2003.

The following year astronomers began carefully tracking the path of another space object. On September 29, 2004, newspaper headlines shouted that a big, speedy asteroid was heading toward planet Earth!

She has never stopped
dreaming
of life forms on other worlds.

Now she is *thinking*
about how life
could come from other worlds.

"I AM FROM EARTH!"

10

It's 6:30 A.M. on September 29, 2004. Adriana Ocampo is peacefully asleep in her house in La Canada, California, a town near JPL.

Meanwhile, astronomers elsewhere on Earth are closely watching an NEO, a near-Earth object that's tumbling through space like a poorly thrown football at rocket speed. Asteroid 4179, known as Toutatis, is headed toward Earth! It's big enough to cause global damage: a fireball, earthquakes, tsunamis, showers of deadly ejecta, and gloomy, ash-filled skies.

Will it hit us—or miss?

You might be thinking that asteroid Toutatis is a dream or a creation of Adriana Ocampo's rich imagination, like the *asteroide* she encountered on the rooftop in Buenos Aires. But this space rock is very real.

So what happens next? Are we doomed? You already know, since you are alive and well and reading this book some time in the future. At 6:35 A.M., the asteroid zooms within a million miles of our planet—a near-miss in space terms. Planet Earth is totally unscathed, and, like Adriana, everyone sleeps easy.

Hundreds of NEOs like Toutatis cross Earth's orbit, and a comet or asteroid will definitely strike Earth again, but Adriana knows there's no point in panicking. "Super-killer" space rocks,

A space artist imagined what Earth would look like from Toutatis *(above)*, a real asteroid that crossed our path in 2004. Adriana continues to scour the globe for signs of past impacts— craters, ejecta, and meteorites (fallen space rocks). Leaving no continent unturned, she traveled all the way to Antarctica in 1999 *(opposite).*

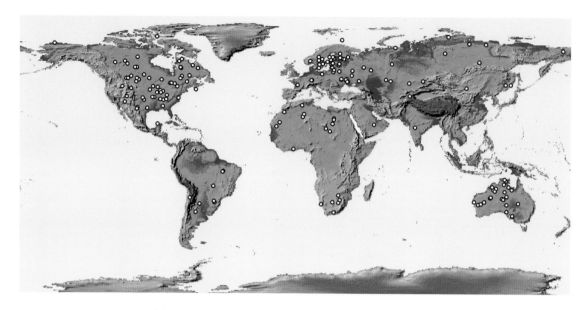

After decades of searching with space-age technology, scientists have identified 172 impact craters on Earth. (This was the count as of November 2004.) They still discover a couple each year, but it's not easy. Most of Earth's craters are badly eroded or buried under younger rock. Very few are as gigantic as Chicxulub, which is good news since it means that "killer" space rocks are rare.

like the kind that made the Crater of Doom, happen only about once every 100 million years.

It's tempting to keep a close telescopic eye on NEOs, but Adriana isn't an astronomer. She explores space by looking down at the ground, not up. Her scientific mind is focused on understanding events of the past, not predicting events of the future.

As Adriana wakes up on that uneventful September morning, she thinks about the present.

~ Not-So-Alien Thoughts

Adriana and Kevin are now divorced, and she has spent the last six years traveling the globe on international space projects. Back home again in California she is returning to a new office, new desk, and new chair at JPL. But it's up to her to find a new project. In that sense, the NASA center is like the Montessori sessions that Adrianita enjoyed with her mother, Teresa, after school. The scientists and engineers can decide what they want to study and for how long. Then they write a proposal, like Post 509's plan for the EMC project, and seek funding for the project from NASA.

What should Adriana propose? As usual she has her fingers in a lot of scientific pies. At the moment her mind is swimming with

International Adventures

In 1998 Adriana proposed a new position for herself at NASA headquarters in Washington, D.C. Her goal was to get more countries involved in exploring space. Her plan was accepted, and as a Program Executive and International Affairs Specialist, she took charge of NASA partnerships with Russia and Russian-speaking countries, Spain, Portugal, and, of course, Latin America. These ambitious projects included five big space missions whose budgets totaled $390 million!

Adriana also helped advise top U.S. leaders, including Vice President Al Gore, about issues in space science. With others she outlined ideas for future missions and ways to work cooperatively with other countries.

In 2002 Adriana moved from Washington to the Netherlands, a country in Europe. She went to join two spacecraft missions —Mars Express and Venus Express— launched by the European Space Agency. After *Mars Express* was on its way, she returned to JPL in the summer of 2004, ready for whatever further adventures space had to offer.

Adriana explained: "When I have been the most successful is when I created my own opportunity, my own position. I have to find my own place."

In 2000 Adriana traveled to the central Asian country of Kazakhstan for the launch of a European Space Agency mission to study the effects of the Sun's energy *(above)*. Two years later she moved to the Netherlands *(below)*.

The Spiritual Side of a Scientist

Adriana Ocampo was once asked how she could believe in God and in science, too. She answered that she didn't think you had to be an atheist—a nonbeliever—to be a good scientist.

"Science and religion are both searches for the truth, but science relies on facts and religion relies on faith," she said. "Those are two different things, and it's not about choosing one or the other."

As a scientist, she can ask, "Is there life beyond Earth?" The answer lies in searching for evidence—making observations, gathering data, experimenting.

As a religious person, Adriana can look at the same question— "Is life out there?"—in a different way. She has faith that the answer is yes. She has believed that we share the universe with other life forms since she was a little girl, stargazing on that rooftop in Buenos Aires. Faith is not a substitute for facts, she says, but rather a way of thinking about our place in the universe.

an idea born on that rooftop in Buenos Aires. She has never stopped dreaming of life forms on other worlds. Now she is thinking about how life could come from other worlds. Could it hitch a ride on an asteroid?

This idea may sound like science fiction, but it's not. It's part of a new and growing field of science called astrobiology. Among many other projects, Adriana is working on a proposal with astrobiologists at JPL's Center for Life Detection.

As a planetary geologist on the team, Adriana's role would be to dig into some tantalizing clues about past life. For example, she found tiny fossils inside the Pook's pebbles from Belize. If the round stones were blasted out of the Chicxulub crater and fell to Earth as part of the ejecta, did melted limestone preserve the fossils? Could impacted rocks on Mars have fossils, too?

Adriana and the astrobiologists will iron out a detailed plan of investigation. Most JPL proposals are turned down, which is one reason that Adriana often works on several projects at a time. The competition for NASA money, team members, and resources is tough. But as usual, Adriana is optimistic that an exciting science adventure will take shape soon.

~Electronic Eyes on Mars

Adriana's mind continues to churn with possibilities, as she pulls into the JPL parking lot on that sunny morning in September when the space rock Toutatis flew uneventfully past our planet. As soon as she walks past security, she spies a family of mule deer relaxing on the grass. The spotted fawn, two months old, was born amid the buildings made of glass and concrete and steel. *The animals are still happy here*, Adriana thinks with a smile.

An artist created this picture of the European Space Agency's *Mars Express* leaving Earth for the red planet in June 2003. Every 26 months Mars and Earth line up favorably in their orbits for the shortest spacecraft flight possible.

In snowy Moscow, Adriana helped NASA hammer out a plan for the United States and Russia to work together on space missions.

She splits the workday between an office in Building 183 and a station across the lawn in Building 264, where she is working on the Mars Odyssey team. This JPL orbiter is helping to pave the way for a human mission to Mars by searching for the presence of water on the surface. A water-rich area would make an excellent landing site for future astronauts.

Adriana thinks about those future pioneers—today's kids and teenagers who are dreaming, as she did, of exploring space. She grew up at the dawn of the Space Age, when the first satellites and astronauts lifted off the surface of Earth. She's both pleased and proud that we earthlings now have more electronic eyes on Mars than ever before. The rovers *Spirit* and *Opportunity* are wheeling around the surface. In addition to *Mars Odyssey*, there are two other orbiters: the European Space Agency's *Mars Express* and JPL's *Mars Global Surveyor*. A fourth spacecraft, the *Mars Reconnaissance Orbiter*, is scheduled for launch in August 2005, with many more to follow.

> Adriana thinks about those future pioneers—today's kids and teenagers who are dreaming, as she did, of exploring space.

Despite all these missions, our exploration of other planets is just getting started, and Adriana feels very lucky to be a part of it. The following week, on October 4, she hatches yet another bright idea for exploring space. The bulb in her brain lights up as she attends the launch of *SpaceShipOne*, the first private space shuttle. With a single non-NASA pilot aboard, it zooms briefly into space, lands, and earns a $10 million prize for its owners.

Adriana thinks, *What about a Mars prize?* People around the world could compete to build nanosatellites—orbiters the size of a fist—and send them to Mars! Adriana imagines a swarm of tiny electronic eyes around the red planet. In her office in Building 183, she adds "Mars prize!" to her list of *Opciónes*, or options.

~ Down-to-Earth Goals

The notebook Adriana began keeping so long ago in Argentina is now a computer file, but there are holdovers: "Astronaut" has been listed for more than 40 years. Adriana applied several times but wasn't chosen. After traveling the globe, though, she has more down-to-earth goals.

She is currently planning a Planetary Society expedition to her home country, Argentina, to explore a KT boundary site. She hopes that investigating KT sites in the Southern Hemisphere will give a better picture of how Chicxulub crater ejecta spread over the globe. At age 49 she is still mixing work and school. Adriana is close to earning her Ph.D. degree, which would finally give her the title

From 1998 to 2002 Adriana lived in Washington, D.C., and worked at NASA headquarters. She was busy but still found time to play tourist in front of the U.S. Capitol building *(below, far left)*. In Washington she could also visit with her sister, Dr. Sonia Ocampo, who has her own psychology business there. Sonia *(below, in blue)* and Adriana also spend time with their mother, Teresa (in the middle), and father, Victor, who are retired.

Mars is too frozen for liquid water to flow, so how did Yuty crater get those "petals" of fluidized ejecta? Was the planet warmer long ago? Or did a crashing space rock melt ice in the ground? Future Martian explorers, robot and human, have much to investigate.

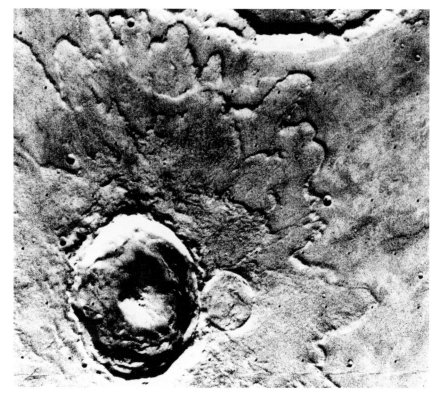

"Dr. Ocampo." More than half the scientists and engineers at JPL have advanced degrees, often earned in their 20s and 30s. Adriana has worked alongside these "doctors" since she was a teenager, but joining space missions meant taking a long and winding path through school.

To earn her final degree, Adriana has to publish more papers based on original research. She has finished her thesis, her main Ph.D. paper, about the crater ejecta discoveries in Belize. There are similar petal-shaped blankets around a few craters on Mars. When future rovers or astronauts investigate those Martian craters, Adriana hopes her data will help explain how the craters got that way.

At the top of her *Opciónes* list, Adriana has written: "Bring space technology to developing countries." This goal is partly due to Sonia's influence. Her big sister has stayed true to a childhood wish to improve the world. As a psychologist, Dr. Sonia Ocampo helps disadvantaged parents raise their children.

Adriana's primary goal also grew out of her international experiences. She helped organize several space conferences for the

United Nations, the group of leaders from different countries who work to solve global problems. Scientists talked about using satellite images and weather data to improve people's lives—by forecasting hurricanes and droughts, for example. Adriana firmly believes that all countries, no matter how poor, should have this kind of technology.

A lifelong goal has been to share her passion for space and geology everywhere she travels. The people she meets often ask her where she is from, perhaps tuning in to her Spanish accent. Adriana Ocampo was born in Colombia, raised in Argentina, became an American citizen, and lived in Europe for a few years. But for a world traveler and space explorer like her, there can be only one answer: "I am from Earth!"

Timeline of Adriana Ocampo's Life

1955   Adriana Cristian Ocampo is born in Colombia, South America, on January 5, joining older sister Sonia. The Ocampo family moves to Buenos Aires, Argentina.

1968   Adriana's father, Victor, emigrates to the United States to look for work.

1969   The Ocampo family settles in Los Angeles, California.

1972   Adriana joins Space Exploration Post 509. In December she watches the launch of *Apollo 17*.

1973   Adriana graduates from South Pasadena High School and starts her career at the Jet Propulsion Laboratory (JPL), part of the National Aeronautics and Space Administration (NASA).

1976   She joins JPL's Viking mission to Mars.

1977   Adriana earns her associate degree in science from Pasadena City College.

1980   Adriana joins the Voyager mission to the gas giant planets. She also becomes an American citizen.

1983   Adriana earns a bachelor's degree in geology from California State University, Los Angeles (CSULA).

1984   She joins the Galileo mission to Jupiter. She also processes images of Earth taken by satellites in orbit around Earth.

1986   On January 28, the space shuttle *Challenger* explodes shortly after liftoff, killing all crew members onboard.

1987   Adriana organizes the first of many planetary sciences workshops through the United Nations and other organizations to promote space exploration in countries around the world.

1988 She identifies a ring of cenotes (sinkholes) in a satellite image as evidence of a buried impact crater in the Yucatán, Mexico.

1989 *Galileo* is launched on October 18. Adriana and archaeologist Kevin Pope marry on November 3.

1990 *Galileo* "discovers life" on Earth. Adriana organizes the first "Space Conference of the Americas," a conference series held through the United Nations.

1991 In Belize, Adriana and Kevin discover ejecta (blasted-out rock) from the Chicxulub crater.

1995 Adriana and Kevin lead the first of four Planetary Society expeditions to Belize to investigate crater ejecta. *Galileo* arrives at Jupiter. Adriana programs the spacecraft to take images of the moon Europa.

1996 Adriana and Kevin lead an expedition to Gubbio, Italy, to explore the first KT (Cretaceous-Tertiary) boundary site used to prove that a massive space rock crashed on Earth 65 million years ago.

1997 Adriana earns a master's degree in geology from California State University, Northridge (CSUN).

1998 She manages international space missions at NASA headquarters in Washington, D.C. She and Kevin separate and later divorce.

2002 Adriana joins the European Space Agency's Mars Express mission, based in the Netherlands. She is named one of *Discover* magazine's "50 Most Important Women in Science."

2004 She returns to JPL and joins the Mars Odyssey mission.

2005 Adriana Ocampo is finishing her Ph.D., working on space missions, planning geology expeditions, and writing proposals for science studies.

About the Author

Lorraine Jean Hopping learned to love science by writing about it. She was the founding editor of Scholastic Inc.'s *SuperScience* magazine for students in grades four to six and has written 30 books for children of all ages, including another biography in this series, *Bone Detective.* She also invented the award-winning Mars 2020 and a dozen other board games published by Aristoplay. Her latest adventure in science is editing Joy Hakim's *Story of Science* book series. Lorraine lives in Ann Arbor, Michigan, with her husband Chris and two cats.

GLOSSARY

This book is about planetary geology. Geologists are scientists who can "read" the features of rocks to learn clues about the past. They think about what events, forces, and conditions shaped the rocks—in short, how the rocks got that way. The term *geology* comes from the Greek word *geo*, meaning "Earth" (the planet) or "earth" (the ground beneath your feet). Besides geology (the study of Earth), there's *geography* (mapping Earth) and *geometry*, which means "Earth measurement." But in this age of space exploration, *geo-* extends far beyond Earth. Planetary geology includes planets, moons, asteroids, comets, and other solid bodies in the universe.

For more information about these words, consult your dictionary.

anomaly: something out of the ordinary—abnormal, irregular, or odd

asteroid: a space rock that doesn't orbit a planet (in which case it would be a moon). Asteroid means "star-shaped," but all except the largest asteroids have irregular shapes, like potatoes. Asteroids are also called minor planets.

breccia: fine-grained stone that contains sharp-edged chunks of rock embedded like nuts and raisins in a cookie

caldera: a large hole on a volcano that looks like an impact crater, but isn't. It can form when a volcano blasts chunks of itself into the air or when an underground chamber of magma (molten rock) empties out, and the ground above it collapses.

cenote: a sinkhole of fresh water. After underground rock is eaten away by acids, the surface sinks into the empty space below and fills with rain and groundwater.

comet: a "dirty space snowball" made of ice, minerals, and gases

Cretaceous period: the time period covering between about 144 million and 65 million years ago. It's the third and last period of the Mesozoic Era ("The Age of Reptiles"). Cretaceous means "chalky," named after sedimentary rock formed during that period.

density: the mass of an object divided by its volume. (Volume is the amount of space it takes up.) The denser the object, the greater its gravitational pull. A gravity map shows the relative densities of rock layers as measured by remote-sensing instruments.

ejecta: rock that's blasted out of the ground by an impact or volcanic eruption. An ejecta blanket is an unbroken sheet of ejecta immediately surrounding an impact crater.

era: in geology, a span of time marked at the start and end by an extreme change in the fossil record, such as a mass extinction. Eras are divided into periods, which are time spans that mark less drastic changes.

impact crater: a hole in the ground made by the crashing of a comet, asteroid, or other object

iridium: a metal element similar to platinum. It's rare on the surface of Earth but common in space objects.

KT boundary: the dividing line between Cretaceous and Tertiary layers of sedimentary rock

limestone: sedimentary rock made of calcium carbonate, the mineral in seashells

magnetometer: remote-sensing instrument for measuring the magnetic properties of rock

meteor or meteorite: a meteor is a space rock that enters Earth's atmosphere and burns up or vaporizes (turns into gas). A meteorite survives the trip through the air and lands on the surface. The Chicxulub impactor wasn't a meteorite because it vaporized on impact.

planitia: low-lying plain. A planum is a high plain, or plateau.

propulsion: the energy that propels, or pushes, an object forward

remote sensing: gathering data from afar through instruments. Your eyes are remote sensors because they collect the light that bounces off objects. Remote-sensing instruments can measure electromagnetic energy (visible light, infrared, X rays, etc.), magnetic fields, thermal energy (heat), gravitational pull, and much more.

robotic spacecraft: space-going vehicles programmed through an onboard computer to act on their own. There's no pilot, and humans don't operate them by remote control from Earth. Instead, mission controllers transmit computer commands by radio to the machines.

satellite: usually a machine that orbits Earth. A spacecraft that orbits another planet is called an orbiter. Moons are natural satellites.

sedimentary rock: layers of soft rock made of sediments (loose bits) that have cemented into solid stone through pressure

sequencing: programming a spacecraft to take a series of images

tektites: round or teardrop-shaped bits of glass that form when melted rock is blasted into the air, cools, and rains down in little pieces

Tertiary period: the time period covering between about 65 million and 1.6 million years ago. It's the original name for the first period of the current era, the Cenozoic Era ("The Age of Mammals"). Geologists now divide that era differently, starting with the Paleogene Period.

Metric Conversion Chart

When you know:	Multiply by:	To convert to:
Inches	2.54	Centimeters
Feet	0.30	Meters
Yards	0.91	Meters
Miles	1.61	Kilometers
Acres	0.40	Hectares
Pounds	0.45	Kilograms
Centimeters	0.39	Inches
Meters	3.28	Feet
Meters	1.09	Yards
Kilometers	0.62	Miles
Hectares	2.47	Acres
Kilograms	2.20	Pounds

FURTHER RESOURCES

Women's Adventures in Science on the Web

Now that you've met Adriana Ocampo and learned all about her work, are you wondering what it would be like to be a planetary geologist? How about a wildlife biologist, a forensic anthropologist, or a robot designer? It's easy to find out. Just visit the *Women's Adventures in Science* Web site at www.iWASwondering.org. There you can live your own exciting science adventure. Play games, enjoy comics, and practice being a scientist. While you're having fun, you'll also get to meet amazing women scientists who are changing our world.

BOOKS

Alvarez, Walter. *T. Rex and the Crater of Doom.* 2nd ed. New York: Vintage, 1998. Scientist Walter Alvarez led a team that discovered evidence of an asteroid or comet impact that resulted in a mass extinction 65 million years ago. This exciting first-person account is hard to put down.

Hartmann, William K. *A Traveler's Guide to Mars: The Mysterious Landscapes of the Red Planet.* New York: Workman Publishing Company, 2003. Practice your planetary geology skills by examining surprisingly detailed photos of the surface of Mars.

Koppes, Steven N. *Killer Rocks from Outer Space: Asteroids, Comets, and Meteorites.* Minneapolis: Carolrhoda Books, 2001. Learn how to recognize meteorites and impact craters and what scientists are doing to defend Earth from future cosmic collisions. The book includes a map of impact sites worldwide and colorful graphics and photos.

Ride, Sally, and Tam O'Shaughnessy. *Exploring Our Solar System.* New York: Crown Publishers, 2003. Take an exciting tour of the nine planets of our solar system with America's first woman astronaut. Get to know Earth and its neighbors, one planet at a time, through fantastic images and cool planetary and spaceflight information.

Verne, Jules. *From the Earth to the Moon* (1865) and *Around the Moon* (1870). These two well-known science fiction stories are available free online (www.literature.org, www.online-literature.com, and other sites) and in many print editions. Check your school or local library.

Zanda, Brigitte, and Monica Rotaru (Editors). *Meteorites: Their Impact on Science and History.* Cambridge: Cambridge University Press, 2001. Find out what space rocks are made of, what happens when they crash, how to "read" an impact crater, and more in this richly illustrated book.

WEB SITES

Jet Propulsion Lab: www.jpl.nasa.gov
Search this site for more information about the Viking, Voyager, Mars Odyssey, and other robotic spacecraft missions.

NASA: www.nasa.gov
This Web site with its many links is jam-packed with lots of interesting word games, cool space facts, and hands-on activities covering a variety of space science topics.

The Planetary Society: www.planetary.org
The online Learning Center is your guide to the solar system, lunar and planetary missions, and how rockets work. Learn firsthand what it's like to be part of an actual space mission, as scientists and engineers share their experiences with the Galileo mission to Jupiter.

Post 509: www.post509.org
The space exploration club that got Adriana fired up about aerospace engineering is still going strong! Even if you don't live in the Pasadena area, you can enter the annual Space Settlement Design Competition (www.spaceset.org). Teams of high school students (ages 14 to 18) create detailed plans for a future space colony.

Space Imagery Center: http://www.lpl.arizona.edu/SIC
For crater photos, facts, and a clickable version of the world map of impact craters, check out this site at the University of Arizona's Lunar and Planetary Laboratory. With the "Earth Impact Effects Program" (www.lpl.arizona.edu/impacteffects/), you can enter the size, speed, density, angle of impact, and other properties of an asteroid or comet and learn what would happen if it crashed on Earth.

SELECTED BIBLIOGRAPHY

In addition to interviews with Adriana Ocampo, her family, friends, and colleagues, the author did extensive reading and research to write this book. Here are some of the sources she consulted.

Alvarez, Walter. *T. Rex and the Crater of Doom.* 2d ed. New York: Vintage, 1998.

Anderson, Charlene. "Europa: Layers of Mystery." *The Planetary Report* (September-October 1998).

Biek, Robert F. *Geologic Maps: What Are You Standing On?* Salt Lake City: Utah Geology Survey, 1999.

Courtillot, Vincent. *Evolutionary Catastrophes: The Science of Mass Extinction.* 2d ed. New York: Cambridge University Press, 2002.

Fischer, Daniel. *Mission Jupiter: The Spectacular Journey of the Galileo Spacecraft.* New York: Copernicus Books, 2001.

Frankel, Charles. *The End of the Dinosaurs: Chicxulub Crater and Mass Extinctions.* New York: Cambridge University Press, 1999.

Godwin, Robert, ed. *Mars: The NASA Mission Reports,* Vol. 2. rev. ed. Burlington, Ontario: Apogee Books, 2004.

Hanlon, Michael. *The Worlds of Galileo: The Inside Story of NASA's Mission to Jupiter.* New York: St. Martin's Press, 2001.

Hartmann, William K. *A Traveler's Guide to Mars: The Mysterious Landscapes of the Red Planet.* New York: Workman Publishing Company, 2003.

Hochman, Gary (producer). *Adriana Ocampo: Space Geologist.* Wonderwise: Women in Science Learning Series multimedia instructional kit. University of Nebraska State Museum and Nebraska 4-H Youth Development.

Jablow, Valerie. "A Tale of Two Rocks." *Smithsonian* (April 1998).

Koppes, Clayton R. *JPL and the American Space Program: A History of the Jet Propulsion Laboratory.* New Haven, CT: Yale University Press, 1982.

Lewis, John S. *Rain of Iron and Ice: The Very Real Threat of Comet and Asteroid Bombardment.* Reading, MA: Addison-Wesley, 1996.

Powell, James L. *Night Comes to the Cretaceous: Dinosaur Extinction and the Tranformation of Modern Geology.* New York: W. H. Freeman, 1998.

Schenk, Paul. "Oceans, Ice Shells, and Life on Europa." *The Planetary Report* (November-December 2002).

Steel, Duncan. *Rogue Asteroids and Doomsday Comets: The Search for the Million Megaton Menace that Threatens Life on Earth.* New York: Wiley, 1995.

INDEX

LIBRARY ADVISORY BOARD

A number of school and public librarians from across the United States kindly reviewed sample designs and text, answered queries about the format of the books, and offered expert advice throughout the book development process. The Joseph Henry Press thanks the following people for their help:

Barry M. Bishop
Director of Library Information Services
Spring Branch Independent School District
Houston, Texas

Danita Eastman
Children's Book Evaluator
County of Los Angeles Public Library
Downey, California

Martha Edmundson
Library Services Coordinator
Denton Public Library
Denton, Texas

Darcy Fair
Children's Services Manager
Yardley-Makefield Branch
Bucks County Free Library
Yardley, Pennsylvania

Kathleen Hanley
School Media Specialist
Commack Road Elementary
Islip, New York

Amy Louttit Johnson
Library Program Specialist
State Library and Archives of Florida
Tallahassee, Florida

Mary Stanton
Juvenile Specialist
Office of Material Selection
Houston Public Library
Houston, Texas

Brenda G. Toole
Supervisor, Instructional Media Services
Panama City, Florida

STUDENT ADVISORY BOARD

The Joseph Henry Press thanks students at the following schools and organizations for their help in critiquing and evaluating the concept for the book series. Their feedback about the design and storytelling was immensely influential in the development of this project.

The Agnes Irwin School, Rosemont, Pennsylvania
La Colina Junior High School, Santa Barbara, California
The Hockaday School, Dallas, Texas
Girl Scouts of Central Maryland, Junior Girl Scout Troop #545
Girl Scouts of Central Maryland, Junior Girl Scout Troop #212

JHP Executive Editor: Stephen Mautner

Series Managing Editor: Terrell D. Smith

Designer: Francesca Moghari

Illustration research: Joan Mathys

Special contributors: Roberta Conlan, Meredith DeSousa, Denton Ebel, Sally Groom, Mary Kalamaras, Emily Kohn, Dorothy Lewis, April Luehmann, Anita Schwartz, Jan Smit, Leanne Sullivan, Aileen Yingst

Graphic design assistance: Michael Dudzik